Sustainable Management for Dams and Waters

William Jobin

Cover Illustration by Fran Eisemann

LEWIS PUBLISHERS
Boca Raton Boston London New York Washington, D.C.

Acquiring Editor:	Dennis McClellan
Project Editor:	Joanne Blake
Marketing Manager:	Arline Massey
Cover design:	Dawn Boyd
PrePress:	Kevin Luong

Library of Congress Cataloging-in-Publication Data

Jobin, William R.
  Sustainable management for dams and waters / William R. Jobin.
    p. cm.
  Includes bibliographical references and index.
  ISBN 1-57444-062-4 (alk. paper)
    1. Dams--New England--Management.   2. Dams--America--Management.
  3. Water-supply--New England--Management.   4. Water-supply--America-
  -Management.   I. Title.
  TC556.5.N48J63   1998
  333.91′15′0974--dc21   98-14705
                                                                              CIP

Preface

This book was written to encourage and educate scientists, resource planners, and engineers who work on water projects to add two dimensions to their understanding of water — the dimensions of ecology and of history. Aquatic and marine ecology encompass the interconnectedness of biology, chemistry, hydrodynamics, and physics of the living waters we dam and divert for our purposes. History illuminates the impact of our damming on related human communities, and gives us a guide on the way to manage the waters so that they serve human needs in the future. These new dimensions require that water projects involve a broad range of people and disciplines; they are no longer the province of a few talented engineers.

This is a book for real-world planners who must be aware of currents in human society and history, as well as currents in two-dimensional models of reservoirs and harbors. It calls for a broad planning approach, linking disciplines. It connects aspects of history, geography, biology, sociology, economics, and politics in planning for management of water resources, and adds a previously neglected focus on the populations of people, other animals, and vegetation that are dependent on the water resources of the Americas.

Long ago, as engineers, we were taught to view water from the narrow, hydrodynamic viewpoint. But water is a medium for life, including our lives — the deep waters of the Great Lakes, the mighty Mississippi River, the cold but productive North Atlantic Ocean, and the crashing Pacific coast. It is much more than moving particles guided by pressure gradients. And history shows us that dam design is a social enterprise, not just an engineering exercise. Political will and organization, hope for the future, egotism, greed, and even dreams are involved; they must not be discounted.

This book is organized to give insight and encouragement to those who would be faithful stewards of our rivers and coastal waters, by following the history of generations of effort to control and use water in the Western Hemisphere. It reflects the slow, steady progress made, especially in this last generation. The book is intended to offer some hope for the generations to come.

This book calls for a break with certain traditional practices, especially difficult for civil engineers. The field of civil engineering is remarkably focused on traditional methods, being closely allied to the arts of the builders, some of which have not changed for centuries. Masonry construction of dams and even bridges involves concepts developed during the Holy Roman Empire. Sanitary engineering, which deals with disposal of human and industrial wastes, uses methods in 1997 such as sedimentation and passive oxidation, which are centuries old. We are constrained by these traditions.

The traditional concepts challenged in this book include the misuse of protected or purified water supplies for purposes other than direct human consumption. The worst examples of this are the construction of urban waste removal systems based on water carriage in sewers, and the haphazard and environmentally disastrous mixing of human and industrial wastes in sewerage systems. The design of urban water supply and sewage disposal systems in the Americas has been based on these concepts for centuries; thus, changing them requires a concerted and intelligent effort.

Because of its broad scope, this book can be used for courses on planning dam and water projects, as well as for courses on aquatic ecology and water resource planning. However, because of the large amount of recent data, it can also serve as a detailed guide to recent water quality conditions in certain parts of the Americas, especially Massachusetts and the island of Puerto Rico in the Caribbean Sea.

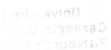

Acknowledgments

If this book serves a useful purpose, it is due to the encouragement of Dennis McClellan of St. Lucie Press and Edward Lewis of Lewis Publishers, who gave me real assistance and good advice some years ago when I first proposed the idea. If the book truly reflects the ecology of the waters of America it is due to the field experience I received under Alan Cooperman, Russell Isaac, Tom McMahon, Al Ferullo, and Jack Elwood of the Massachusetts Department of Environmental Protection. If the book embodies any of the philosophy of Native Americans it is due to the wise teachings of the late spiritual leader of the Wampanoag People, Slow Turtle. If the book shows the joy I find in God's Creation it is due to the insight provided into the quiet and intrinsic beauty of Neponset Reservoir given to me by Sheila Warner, Frank Bonacorso, Neil Kaiser, and Fran Eisemann of Foxboro, Massachusetts — as well as Phyllis Whitehead and Marc Marcussen and their students in the Foxboro schools. The cover drawing of the salmon and the most creative graphics are by Fran Eisemann and Laura Jobin. If the book is readable and flows logically, it is due to the help from Jane Kinney, Environmental Editor of CRC Press in New York. And if the text and graphics are well composed and if there are few errors, it is due to the creativity and thoroughness of Joanne Blake of CRC Press in Florida. The faults and inadvertent errors in this book, however, are due to my own enthusiasm and inexperience.

Several millenia in the past, the ancient Pharaohs were devising structures on the Nile River to divert water for agriculture. Traces of ancient dams from Sri Lanka, China, Spain, and the high mesas of Colorado indicate that dam building was a universal human activity since the dawn of our civilizations. In the Americas, the Connecticut River of New England is only 400 miles long; yet its drainage basin contains over 700 dams, most of them constructed in the colonial era three centuries ago. As important as this construction activity was in the past, rehabilitation and operation of these dams will be even more essential in the future, and we must learn to do it right; thus this book.

We dam rivers because water is so fundamental to our lives. Water is a basic element of Creation — like earth, air, and fire. We are drawn to the drama of rivers, the peace of the lakes, and the vastness of the ocean by a primeval attraction. We sense a mystical connection with our deepest nature in the ebb and flow of these waters.

This book is organized around hydrologic drainage basins, partly because these river basins or watersheds are the basic geographic unit for dam design. But the water in the rivers is also a medium for life and has an eternal dimension, being continually renewed in the global hydrologic cycle. It is not surprising then that the church that led me to life has recently reorganized according to river basins, recognizing watersheds as evidence of God's Creation.

In their explanation of this new alignment, I was heartened that the church elders also recognized the ecological importance of our management and use of water. This is clear in the reason given for their new hydrologic emphasis:

> It is difficult for us to recognize the ecological relevance that watersheds play in our lives, because we have developed a "faucet mentality." We no longer know where our water comes from, or where it goes once it has disappeared down the drain... We have lost sight of its origin and value. We've grown to believe it is inexhaustible... When we understand that the water we squander and pollute today may have been the same water used to baptize Christ, then hopefully our relationship with it will change... This new perspective revitalizes our sense of community...

The Episcopal Times, Boston, Massachusetts, April 1996

Their logic lays the foundation for the ultimate theme of this book: that the restoration of our rivers and oceans toward their primeval ecology will best be accomplished by a reverent community, defined by its common location in a hydrologic watershed.

The people of the Episcopal Church in eastern Massachusetts, especially those of St. Mark's Church in Foxboro, have thus helped me to see the beauty in the waters of God's Creation as I pondered the past and future of the lakes, rivers, and oceans of New England and the rest of the Americas. More than that, they baptized my children, nurtured me in times of trial, and celebrated with me in the joys of marriage, birth, and adoption. For over 30 years, they have been my home and family. I dedicate this book to them, with the hope that as they form new communities by watersheds, they will partake of healing in the waters.

Contents

SECTION III
Our Hope for the Future

CHAPTER 5
Citizen Watershed Associations

CHAPTER 6
Dam Designers and Water Planners

Sustainable Use of Water Resources in the Americas

PROLOGUE

Turtles, According to Nakos in the 19th Century, a Chief of the Arapaho People from the Black Hills of North America

Long ago, all was water, and the turtle went to the bottom of the water and brought up a bit of clay. Out of the clay the world was made, and thereafter the turtle became a symbol of the earth. The ridge on his back is the mountain line, and the marks are streams and rivers. He himself is like a bit of land in the midst of water. (From *The Indians' Book,* 1987, edited by Natalie Curtis, Bonanza Books, New York.)

Turtles, According to the Massachusetts Audubon Society

Of 15 native turtles, those in danger of disappearing forever from Massachusetts waters are the green turtle, the ridley turtle, the hawksbill turtle, the muhlenbergs turtle, and the red-bellied turtle. (From J.D. Lazell, *Reptiles and Amphibians,* 1972, Massachusetts Audubon Society, Boston.)

For 200 million years, reptiles and amphibians dominated the Earth, until about 100 million years ago when the Rocky Mountains rose, and we mammals began to establish our rule. Will we soon exterminate the last of these ancient turtles?

CHAPTER 1

Introduction

The subject of this book is the rehabilitation and sustainable management of water resources in the Americas, especially the planning and modification of dams, rivers, lakes, sewers, harbors, and domestic water supplies. The last three centuries of experience with water projects in North and South America are examined in the book. From this analysis, new guidelines are derived for planners of dam and water projects that will lead them toward sustainable designs.

The guidelines apply to planning and design of new dams and new canals, and also to rehabilitation of existing structures and systems. They include a refocusing on correction of river, lake, and harbor pollution problems by turning the construction-oriented and negative regulatory programs of our generation into a positive and self-sustaining program for coming generations of new Americans.

1.1 RESTORATION OF FISHERIES

This positive approach to restoration of our waters emphasizes the goal of reestablishing primeval fish populations as a renewable and sustainable food supply. Focusing on restoration of primeval fisheries is another step forward from current water quality goals that, in the 1960s, established the chemical conditions necessary for sensitive stages in the life of fish. But chemical water quality is not enough. Aquatic life requires more.

Restoration of primeval fish and marine mammal populations is proposed as a new focus for environmental programs related to water, because these animals are increasingly important to diversify and strengthen our food supply, and because the specific conditions required for their health are conditions that will also restore the ecology of our waters.

Our economies would also benefit from other aspects of the restoration of natural fish populations: the restored production of fur and fiber that would follow from sustainable management and harvesting of revitalized wetlands. And our souls would benefit from enhanced recreation.

1.2 HISTORY AND SUSTAINABLE EXAMPLES

Underlying this book and the engineering and planning guidelines that it develops are two other important assumptions. The first assumption is that engineering schools must add the study of history to their curricula for designers and planners of dams, canals, sewers, and water projects, in order to avoid repetition of the great water engineering blunders of the past century.

The second assumption is that water resource planners, including dam engineers and environmental scientists, must learn from examples of sustainable projects provided by our diverse American community. Eager young engineers from the pampas of Argentina, the Rocky Mountains of North America, and the coast of New England can learn important design concepts from experienced alfalfa farmers of central Mexico, descendants of the conquistadores and the ancient Mayans. Crusty New England organizers of watershed associations can learn about sustainable wild-life programs from Wild Westerners, and about fisheries restoration from Klamath fisherfolk and the new generation of fisheries planners in the Columbia River Basin of the Pacific Northwest. Conservation of water and reuse of human sewage around the North American cities of Denver and Boston could be improved by following examples from Mexico City. Citizens of water-scarce areas in Colorado and Nevada can benefit a great deal from the decades of political experience of the watershed groups of Massachusetts. And the enthusiastic dam planners of Brazil and Uruguay might acquire immunity from the "concrete fever" of the U.S. Bureau of Reclamation and Corps of Engineers, and thus be protected against repeating their mistakes. We are a diverse community in one hemisphere; we need to understand each other's history of mistakes and successes.

There are also obvious needs for dam engineers to understand the creatures which live in the rivers and great seas that engineers divert, obstruct, and connect. Turtles, fish, or algae; mollusks, otters, muskrats, and beavers; snails and whales should all be appreciated by water planners and engineers. Thus, engineers need to listen to biologists, and aquatic ecologists must learn the language of dam builders. Our diversity of approaches to protecting our environment is a strength if we learn to speak in each other's language. This book is an attempt to start the conversation.

At the beginning of the 20th Century, dams were popular because they promised abundant power, wealth, and food security. But now, at the beginning of the 21st Century, dams are extremely unpopular because of their negative impacts on the ecology of our rivers and coastal waters, and because many of the dams are now a century old, and failing. However, our food requirements continue to grow while our food sources in the rivers and oceans disappear. We must restore the original productivity of our rivers and oceans because we need the aquatic and marine food. At the same time, there is competition from farmers growing grain who need more water for irrigation.

The conditions of our waterways as well as the diminishing supplies of food from rivers and coastlines are causing a dramatic change in our future way of living, to its detriment. To correct these conditions, we will have to build more dams and rehabilitate existing dams. Thus, we need a new, comprehensive, and enlightened approach to dam design.

Figure 1.1 First dam at mouth of Neponset River in Milton, Massachusetts, and the abandoned Baker Chocolate Factory. One can easily locate rivers in Massachusetts by looking for the tall smokestacks of abandoned mills.

There are many dams in New England in need of rehabilitation, such as the Adams Street Dam at the mouth of the Neponset River near Boston Harbor (Figure 1.1). This simple, low dam, and its predecessor constructed in the early years of American independence, was built in a single-minded scheme to provide water for mills. The first version was used to power the grist mill of Josiah Stoughton, to grind corn. The final, existing version was used to supply process water for the Baker Chocolate Factory. Unfortunately, the dam, since its version was constructed at the end of the 18th Century, has also destroyed the migratory fish populations that originally attracted colonial settlers to the falls in Canton and Norwood, upstream. For 200 years, this dam has blocked passage of herring, menhaden, shad, salmon, and perhaps even sturgeon. The now useless dam could easily be breached as the first basic step in a move to restore the Neponset River to its primeval condition.

It isn't as if people haven't tried to restore the fisheries on the Neponset River before. Some of my favorite stories about the Neponset River illustrate the persistence of fishermen trying to get rid of the Adams Street Dam. In the mid-1700s when the first version of the dam was built, residents of Norwood and Canton tried legal measures in the colonial court system, but were outmaneuvered by the dam builders, including Paul Revere. Much later, when the more permanent dam structure was being built, fishermen from Canton tried to blow it up with a wagonload of dynamite. Unfortunately, they were stopped by the local militia.

We also need a new approach to the disposal of human and industrial wastes — which clearly has not yet been developed as a sustainable science. Repeated failures of urban sewage systems during the past three centuries — such as the one around Boston Harbor in New England — should be a warning that there are some basic flaws in our design principles. As their costs for safe water and sewage disposal skyrocket, the people of Boston will soon be acutely aware of these flaws.

This book describes the ecological requirements for sustainable dams and waste-water systems. It also develops the economic and ecological justification for restoring American rivers and estuaries to their original wholesome ecology. Our waters must not only be suitable for fishing, swimming, and water supply, but we should actually be using them for those purposes. We do not.

The material covered in this book includes operational aspects of water resource management, as well as considerations for watershed associations of citizens to participate in financing and enforcing restoration programs. The inefficiency of state and federal toxic waste programs when applied to industries that contaminate our water is illustrated, and suggestions are made for more fruitful approaches.

A planning element is developed in some detail regarding the need for economic evaluation of damaged ecology of contaminated waters, based on the potential productivity and harvesting of restored fisheries, on the economic value of recreational potential, and on impacts on waterside property values. New data are presented on losses in property values from industrially contaminated or eutrophic lakes in New England. This information should help planners in conducting more precise economic analyses of their projects.

1.3 REVERENCE FOR CREATION

Water is a medium for life; life that is a miracle given by our Creator. Thus, water and life are to be revered, not exploited. From the story of Creation in the scriptures of the Hebrew people — as in many other Creation stories — the message is clear that the Creator put this life in the waters. He called the dry land Earth; and the waters that were gathered together, God called Seas. And later in this story, God created the great sea monsters and every living creature that moves, with which the waters swarm, according to their kinds, and every winged bird according to its kind. And God saw that it was good. And God blessed these creatures of the waters and sky saying,

Be fruitful and multiply and fill the waters in the seas, and let birds multiply on the earth

— The Book of Genesis

In 1995, 25 years after the first Earth Day, the U.S. Environmental Protection Agency (EPA) reported that 40% of the nation's rivers and lakes were not fit for drinking, fishing, or swimming. The previous year, the agency issued 1000 warnings against eating fish from chemically contaminated waters. In 1996, the Nature Conservancy reported that in North America, 67% of mussels, 64% of crayfish, 36% of fish, and 35% of amphibians were either in jeopardy or already gone. Ten fish species have disappeared in the last decade in North America. This dismal situation demonstrates our lack of reverence.

CHAPTER 2

Sustainable Societies and Water Resources

Many of our ecological principles were understood by ancient cultures. In the Hebrew scriptures, from the wisdom of King Solomon or his scribes, we are told that:

*A generation goes, and a generation comes,
but the earth remains forever.
The sun rises and the sun goes down,
and hastens to the place where it rises.
The wind goes to the south
and goes round to the north;
round and round goes the wind
and on its circuits the wind returns.
All streams run to the sea,
but the sea is not full:
to the place where the streams flow,
there they flow again.*

Despite the lyric comfort of these words, and their demonstration of the early and profound understanding in Western Civilization of the hydrologic cycle and the renewable nature of Creation, there are some ecological catastrophes caused by this same civilization in this century that make us want to scream in rage and despair. The near annihilations of the right whale from the North Atlantic Ocean and of the great blue whales from all the oceans of the globe are perhaps the worst. These magnificent creatures supplied the whalers of New England and seafaring countries with a bountiful living, until the whaling industry improved its technology and perfected their mass destruction. Because of the mystical beauty and global range of these enormous cetaceans, their loss was perhaps even more tragic than the annihilation from North America of the herds of buffalo.

It is hard to imagine how an intelligent society could permit whaling — an activity at the base of the 19th Century New England character and economy — to disappear by the 20th Century without earnest thought and energy given to its preservation. The hardy New England sailing captains and their polyglot crews were the legendary core of the coastal culture of this seafaring region. Captain Ahab of New Bedford personified the intimate and almost obsessive relation between the crusty New Englanders and the whales, as described in the ornate and powerful novel *Moby Dick,* by Herman Melville. The loss of the great whale populations was a tragedy for our global ecology, besides devastating coastal cities that have never recovered.

A second catastrophe for New England occurred more recently, but only the desperate and now unemployed fishermen noticed. The pelagic fishing industry, centered around Georges Bank off Massachusetts and around the Grand Banks off Canada, died in 1995 because of ecological ignorance. A fundamental economic activity that nurtured and formed the society and culture of the entire northeastern coast of the continent for three centuries, suddenly died.

The collapse of the whaling industry, the pelagic fisheries, and even the freshwater fisheries were also accompanied by the disappearance of the New England farm. The thin topsoil and erratic climate was finally too much for these hardy farmers, although they had endured since the 17th Century under all sorts of adversities. By the end of the 20th Century, only cranberries and apples could be profitably grown in Massachusetts, and the cranberries needed irrigation. Corn and alfalfa could be irrigated, but how could New England corn compete with crops from the Mississippi Valley? With the loss of the New England farm, another component of the region's character slipped into oblivion. What would Robert Frost say of the untended fields and stone fences that could no longer make good neighbors? What will it take for people to notice the losses?

As a nation, the U.S. did not even blink at the 1995 collapse of the Atlantic fishing industry, although a few well-meaning politicians offered to pay compensation for the idled fishing boats. They must have felt some involvement as it was their effort in Congress to equip these boats with more effective fishing gear that hastened the end of the fish. The purpose of U.S. government grants to improve fishing boats in the 1980s was to help the fishermen increase their declining catches. However, what was needed was an appreciation that the fish stocks were finite, that only a lower rate of harvesting was sustainable, and that the fish habitats had to be restored.

By 1989, all oceanic fisheries around the globe were being fished at or beyond their capacity. Of the world's 15 leading oceanic fisheries, 13 were in decline.

But no one paid any attention because New Englanders had become inured to the disappearance of freshwater fisheries from their rivers during the previous century, to the loss of their farms in the early part of this century, and to the more recent loss of clams, lobsters, and crabs from their bays and harbors, just as the people of the Wild West had accepted the loss of the buffalo. With their technological marvels, they blithely assumed that they would always find ways to harvest or produce food somewhere else.

In our ignorance, we are destroying the magnificent building blocks of Creation, the beautiful fruit of millennia of evolution. There has to be a better way.

2.1 HISTORICAL PERSPECTIVE ON DAM BUILDERS

In many cases, the ecological destruction of fish and other creatures in our waters is the byproduct of an innate drive of mankind to tame these waters by building dams, canals, and conduits. Despite its ancient legacy of tradition and artisanal practices, the building of dams and diversion of waters has started to evolve into a new profession whose shape we can only dimly see in the future. The profession has undergone a long evolution, and now it is in the midst of revolution.

As children, we all fashioned small dams of mud or stones in the rain, diverting torrents and blocking drains; so we have an inherent feel for the living and intriguing behavior of water. The Pharaohs and their engineers probably started with this same experience, then began to tame the Nile River by modest steps, adding blocks of stone to slightly raise the river level upstream, followed by bolder attempts to block off the entire flow. Undoubtedly, the failures were many; we have only a few reminders of their successes. The ability to quarry, float, and place large pieces of stone was probably the major concern of these early water engineers. The Nile River was a large, regular, and well-understood factor; its level had only to be raised a small amount to provide a great benefit for irrigated agriculture.

Much later, the mill builders of the British Isles combined low masonry dams with wooden waterwheels and mills. They used hand tools and animal strength to build devices that captured small streams for domestic power, increasing the size of their projects very gradually, as experience permitted. Water-powered mills ground corn and cut logs. In these early stages, the dam builders focused on their tools and building materials. Tools put the limits on the size of their endeavors. The building materials such as wood also limited the pressures of water that could be harnessed, and the speed of the spinning wheels. These materials remained the same over centuries.

When the British colonists migrated to the Americas, they brought their dam-building skills with them, but the tools and materials changed little. Fortunately, the geography of the east coast of North America was similar to that of the British Isles, so the myriad dams on the rocky rills of New England were easily patterned after those back home (Figure 2.1). In the colonial era and the first century after Independence, dam building was a learned trade, not far removed from carpentry, barrel making, and masonry.

2.1.1 Hydroelectric Power

There was eventually a change from wood and rock to metal and concrete construction at the end of the 19th Century when the first hydroelectric project was constructed in the San Juan Range of the Rocky Mountains at Ames, Colorado. This project provided electrical power for the gold and silver mines being opened in the area. The use of metal allowed fine vanes on the turbines, with curvatures adapted to the flow of the water. Metal bearings allowed much faster spinning of the shafts. Hydraulics laboratories were established to perfect these vanes and speeds, and the technology rapidly changed to utilize even higher pressures and faster velocities of flow. Many civil engineering schools established hydraulics laboratories to test improvements in hydraulic equipment and study theories of flow. Power production increased from horsepower to megawatts in a few short decades. An intense interest in hydraulics developed among civil engineers. Mathematical analyses soon lifted hydraulics to the stature of a science, funded by the drive for more power and bigger turbines.

Figure 2.1   Factory Pond Dam in Canton, Massachusetts. This low masonry dam on the East Branch of the Neponset River was built during the time of Paul Revere, by craftsmen using the basic masonry techniques perfected by the Romans. Similar handiwork can be seen in the simple masonry arches upstream that allow the river to pass under the Canton Viaduct.

With the increased technology of the two World Wars, engineers started on immense dams and irrigation systems in the mountainous and dry western United States. The high dams were actually simple structures, however, using the massive weight of concrete to resist the push of the water. These structures were usually triangular in section, not too different from the pyramids of the Pharaohs.

As dams became higher, they required more careful design. At some point, the amount of concrete required for these huge triangular prisms became prohibitive, and mathematicians began to explore ways to use arches, buttresses, and cantilevers to resist the water pressure, instead of the balance of gravity on the massive masonry blocks (Figure 2.2). French engineers perfected the arch dam almost as an art form, even curving it vertically as well as horizontally. American engineers were not so bold, however, and were further intimidated by the few French dams that blew out at considerable loss in lives downstream. Before computers, the calculation of the shape for these eggshell structures was a daunting task, dominating the attention of many of the best engineers.

During the middle of the 20th Century, roughly the half-century before Earth Day 1970, design and building of dams and irrigation systems were tasks for people with mathematical and engineering skills because of the complexity of new design approaches. The most able and talented of these rose to the top in large water agencies and became administrators and agency leaders; but they were still engineers who

Figure 2.2 Jiguey Dam in the Dominican Republic. This small but relatively high dam, shown under construction in 1990 about 100 miles west of Santo Domingo, represents a compromise between traditional practice and mathematical daring. It is a mass of concrete for safety reasons, but with a modest curvature, to take some advantage from the almost vertical rock walls of the Nizao River.

talked in hydraulic language and thought in equations. Their reactions to social problems created by their projects, such as the displacement of Native Americans from grazing lands upstream of a proposed reservoir, depended on their own personal feelings about Native Americans — probably gained mostly from Hollywood movies about the Wild West. Their reactions to the elimination of ancient salmon runs on the tumbling waters of rivers in the Northwest probably depended on whether they liked to fish — and they probably did not.

The education of civil engineers and dam builders during this period was rigorous but tightly focused. After a groundwork of mathematics, physics, and chemistry, the student was led through the derivation of flow equations, static and dynamic structural analysis, engineering economics, properties and strengths of building materials, construction methods and practices, project design procedures, as well as current design practices in a specialty field such as dams, highways, buildings, canals, or structures.

The engineers and planners relied on economic analyses of costs and benefits to evaluate the financial desirability of a project, or the comparative advantage of variations in its design. Using economics, it was easy to add up figures on each side of the balance sheet and come out with tidy sums in favor or against. There were several variations in the process, depending on interest rates or opportunity costs and the assumed length of project lives. From these numbers, the engineers could calculate cost-effectiveness ratios, cost-benefit ratios, and other seemingly precise indicators.

However, in conducting an economic analysis, it was also necessary to omit many aspects that could not be quantified so easily. It was customary to simply list these as intangibles; for example, esthetics, environment, recreation, or social impacts. Omission of these intangible items was a major problem with many project justifications. Because they could not be measured in dollars, they were often dropped from consideration — until recently.

2.1.2 Recent Major Changes

Two important events in America forced a change in the insular view of the dam engineers: the Civil Rights Movement of the 1960s, and Earth Day — the start of the Green Movement in 1970 — a movement that has just ended. When minority groups in America started to exert political power and demand recognition, it became much more difficult to flood out ancient tribal lands along the Missouri River. When the Green Movement fostered the Endangered Species Act and Environmental Impact Assessments, dam design took on a new flavor; small fish and endangered ecologies suddenly came off the list of intangible items. These events are very recent; their impacts are uneven in different parts of the country and different sectors of our society, but the impacts are unmistakable.

At present, there is usually a socioeconomic and environmental assessment of large water projects before their design is completed and their construction authorized. On the governmental level, this has brought in many new disciplines to the complex process of project design. In an unusually foresighted move, this interdisciplinary process of environmental assessment was instituted by the U.S. Department of the Interior for the existing Glen Canyon Dam on the Colorado River in Arizona (Figure 2.3). The purpose of the post-construction assessment was to evaluate new and environmentally friendly operating rules for the dam when its full complement of turbines is installed in the near future.

Another side of this new process is public participation, usually designed as part of socioeconomic studies, but often a key activity related to protection of the environment. The birth of public interest groups such as the Sierra Club, the Environmental Defense Fund, and myriad watershed and wildlife organizations has brought voices into the discussion that were never heard in the past.

Most recently, as the irrepressible pressures of population growth have caused famine in Africa and Asia because of floods and droughts, the concept of sustainable societies has emerged, implying the need for sustainable supplies of food and water that will always match the number of people. Agriculture, fisheries, and herding, and especially water management, are thus being reevaluated in terms of the distant future, rather than the traditional analyses for short lifespans of a few decades.

2.1.3 Water Planners in the Future

The water planner of the future must therefore be able to handle many more concepts than just flow equations, turbine efficiencies, and flood frequencies. What is perhaps most difficult for planners and especially engineers with traditional educations, is that

Figure 2.3 Glen Canyon Dam was constructed across the Colorado River in 1963. This huge arch dam created Lake Powell, which stretches back into southern Utah.

few of these new concepts are subject to mathematical analyses. One cannot derive the importance of the grazing rights of the Mandan people along the Missouri River in the Dakotas from an analysis of flowing particles in the river channel. One cannot evaluate the importance of the loss of downstream wildlife habitat by an equation that balances the forces on a stationary object.

Forming a team of people with backgrounds in all of these various disciplines is one way of dealing with the complexity of the issues in water planning. Nonetheless, there is usually a team leader, or a key decision-maker, who must integrate the opinions and make a final decision. What should this person use as the basis for decisions on an important water project?

There may no longer be an acceptable, codified set of rules and equations for these decisions. In contrast to the past when engineers could compare calculations on flood heights and reservoir volumes using accepted equations, it is no longer possible to appeal to such professionally blessed standards. Economic analyses have been shown to lack the universality to make comparisons that satisfy the larger aspects of water diversion and management. In fact, the cost-benefit analyses developed a reputation for distortion and deceit by project proponents.

To plan water projects of the future, one needs new guidelines for decision-making, based more on concepts, philosophies, and world views, than solely on principles derived from mathematics. These planning concepts now include sustainability, preservation of species and complex ecologies, democracy, and even justice. Another approach would be to look for the principles found in traditional societies that have shown themselves to be sustainable, such as the original Americans who populated the American continent for millennia without destroying its resources.

2.2 SUSTAINABLE CULTURES

If water planners of the future would look to enduring societies for principles of planning, some clues about these sustainable lifestyles can be gleaned from a study of the teachings of the remnant communities of the original Americans — people who lived in harmony with the ecology of this continent for millennia. The woodland inhabitants of the northeast coast of the continent, the nomadic and hunting cultures of the Great Plains in the Mississippi River Valley, the diverse societies of fisherfolk in the Pacific Northwest, and the Mayan and Aztec builders of South and Central America lived in ecological harmony with their environments, fishing, hunting, and harvesting, supplemented by sustainable agriculture.

Another key aspect of this original American culture that our immigrant society has already adopted is the strong element of democracy and freedom of expression. Much of the current North American style of governance, especially New England Town Meetings and the rules for debate in the subsequent U.S. federal and state congresses, were copied from the democracies of the Iroquois Confederation, as well as the Algonkian culture that covered most of the temperate zone of North America. The principles of equal speaking opportunities for all, development of community consensus, political rights for women, and reverence for wisdom of elders were copied from principles of the Iroquois Federation of Northeastern America. A primary and enduring difference between American congresses and European parliaments is the Iroquois contribution to public discussion. Everyone has a right to be heard. The U.S. Congress is even now ashamedly reminding itself that public discourse should be civil, compared to the British or Europeans who have allowed interruptions and vilification of speakers since their first parliaments of centuries past.

Exposure to the knowledge and understanding of the original Americans in the ways of forest creatures, fishing techniques, and even locally suitable agriculture amazed the Puritans arriving on the shores of North America. Sharing of this knowledge by the Wampanoag, Massachusia, and Narragansett peoples was probably the key factor in survival of the Puritans during their first few years on American soil. Although the Puritans were well organized militarily and had sufficient metal swords and blunderbusses to overcome any resistance to their expansion, they were ignorant of the ways of the deer and wild turkey in the forests, and those of the herring, smelt, and menhaden in the coastal inlets. Without the help of the local people, the Puritans would have died of starvation.

The immigrants gladly accepted the advice of the original Americans when they saw the advantages of their hunting and agricultural methods. Important social aspects of the original Americans were also copied. British and European dominance by royal families and the rowdy behavior in their crude parliaments were eventually left behind by the immigrants, when they saw the advantages in the governing systems of the Iroquois Confederation and other Algonkian groups.

There is an intriguing question regarding the clash between the original Americans and the immigrant and industrial society that washed over them in the 17th Century. How can we explain the technological and military dominance of the immigrants, if the society of the original Americans was founded on such strong and sustainable principles? The answer may be quite relevant, and related to water resource management and irrigation.

One explanation for the technical dominance of the immigrants may be the technological influence on Europeans of the ancient irrigation societies in the Fertile Crescent of the Eastern Mediterranean region. The primeval Americans did not have this technology.

Although the cultures of the original Americans were ahead of the immigrant cultures in terms of government and sustainable lifestyle, it was the scientific and technological advantages enjoyed by the immigrants that ensured their eventual dominance. These advantages were derived from the great hydraulic civilizations that developed along the Nile, Euphrates, and Tigris Rivers because of the highly regular annual floods, perfectly timed with the growing season. Thus, the Pharaohs of the Nile developed crop surpluses that allowed for free time to educate their children, for leisurely studies by educated scientists, and for study and even measurements of the land, the river, the sun, and the stars.

Isolated by deserts from marauding tribes and aided by an extremely dry climate, the Egyptians could save grain from good years to avoid famine in lean years. On this basis, they had the stability to develop geometry, astronomy, calendars, land surveying, and basic hydraulics, eventually providing the Middle East and Europe with science and technology.

The kingdoms of the Nile and Mesopotamia did not last, however, despite their technological prowess. They only lasted a thousand years or so, due to invasions from other kingdoms that developed transport to overcome the Egyptian's desert and sea barriers.

The history of the "hydraulic civilizations" contrasts markedly with that of the North Americans. Although a portion of the peoples inhabiting the Mississippi River Valley in North America had unusually favorable topsoil, hydrology, and seasons, these people never had the necessary isolation to protect them from raiders, so surplus grain could not be accumulated. Every tribe was exposed to raids by neighboring tribes. Wealth, leisure, and the time for education never accumulated, as they did in the ancient kingdoms of Babylonia and Egypt.

In the primeval forests of northeastern America, there was even less favorable climate and topsoil. Easy travel by water made isolation of kingdoms even more difficult. So, neither food nor wealth accumulated. People lived by harvesting the natural resources of the land, water, and sea. Thus, their pharmacopeia never developed sufficiently to reduce death rates, the process that had set off expansion of populations in Europe. Science never developed, nor did the technology for warfare. The resulting characteristics of the original American societies of good government, harmony with nature, and love of the land — as valuable as they are — were not enough to overcome the onslaught of gunpowder and metal swords of the Europeans.

Some of the more notable characteristics of the strong democratic traditions of the Algonkians and other original American cultures included reservation of powers by the people with few powers given to the elected leaders, and a requirement of consensus for making war. This was the opposite of the European traditions, which gave immense power to hereditary rulers, especially the ability to declare war. The only European government that has traditions similar to the Algonkians is Switzerland, a country that has endured under the same constitution for over 700 years, seldom making war on its neighbors because its presidency has always been weak.

For centuries after the initial settlement of the North American continent by Europeans, it was the stratified nature of the warring European societies that drove the immense migration toward the Americas. The immigrants looked for a land where opportunity and a voice in government were not hereditary, nor reserved for privileged families. This contribution of the original Americans to our current societies may be a key factor in developing sustainable institutions to protect our water resources in the future.

2.3 STEADY DESTRUCTION OF NATURAL RESOURCES

Protection of water resources is made very difficult by the inexorable pressures of growing human populations. Biologists know very well that populations of many organisms suffer catastrophic collapses after periods of exponential growth, because overpopulation causes both an escalating contamination of the environment with waste products, and also a rapid depletion of water and food supplies. Thus, population surges are followed by crashes because of famine and environmentally transmitted diseases.

But biologists also know that technological societies of human beings have not followed the patterns of boom and bust suffered by fruit flies, weevils, mice, or deer. Instead, we try to limit our population growth, we increase our food production, and we try to limit the contamination of our environment. Thus, populations of industrial societies have maintained relatively high growth rates for centuries. The parallel stability and accumulation of wealth and knowledge have allowed increasingly rapid development of science and technology, but have also produced enormous populations. The population of Mexico City is about 20 million people.

2.3.1 Overpopulation in Industrial America

Although able to maintain themselves, the industries and great metropolitan areas of these growing industrial societies have nonetheless seriously damaged the surrounding natural resources, water quality, esthetic appeal, and property values — formerly thought of as intangible items by civil engineers and water planners. The principal sources of this damage are unsustainable water and wastewater systems, hazardous wastes, mountains of trash, and crudely designed and eventually abandoned dams. The practice of flawed water engineering has proceeded for over three centuries along the east coast of North America, and in many other countries in the Americas. More recently, the problem has been compounded by the increased production of agricultural, urban, and transportation wastes that wash over the land and seep into the rivers and oceans. This diffuse waste material is most dense in and around the large population agglomerations made possible by modern technology.

I see the New England coastal corridor from Chesapeake Bay to the Gulf of Maine as a sickening example of this decay and contamination, exceeded only by the industrial degradation around lower Lake Michigan and the other Great Lakes. Highway travel into these areas of urban and industrial blight is a shocking and even nauseating experience for unsuspecting travellers, especially those from rural areas. Insane traffic densities, swarming of vehicles around major cities, and interminable waiting on highways are common, as crowds escape from the cities to seek refreshment in ocean waters or in northern woods. These aggravations are hallmarks of the intensely over-populated regions of industrial societies. People grow accustomed to this insanity.

2.3.2 Loss of Food Supplies from Rivers and Oceans

Our world population reached 1 billion in 1830, the product of growth over millennia. To reach the second billion took only 100 years, to 1930. By 1990, our world population jumped to 5 billion, and the next billion will appear in just 11 years. In a manner that compounds the problems of the coastal New England cities, the global human population is now entering an alarming phase; alarming because food reserves are disappearing. In 1990, world grain reserves exceeded the amount needed to supply the world for 100 days. Some 6 years later, the reserve supply was only enough for 49 days, the lowest on record.

In the past century, one easy source for the additional food needed to feed these increased billions was the open sea. However, our technological capacity to expand the rate of harvest of naturally occurring seafoods such as fish and whales has overtaken our scientific understanding of the consequences of such rapacious fishing. Thus, we have depleted or destroyed most of our aquatic and marine food supplies, and are being forced of late to depend almost entirely on increases in agricultural production of grains for additional food.

The destruction of these marine food stocks is evidence of our lack of appreciation for not only their enormous value, but also for the wonder that they represent, the fruit of an amazing Creation.

It is He who has subjected to you the ocean, so that you may eat of its fresh fish and bring up from its depths ornaments to adorn yourself with. Behold the ships ploughing their course through its waters. All this He has created, that you may seek His bounty and render thanks to Him...If you reckoned up God's blessings you could not count them...

The Surah of the Bee
The Holy Quran

The loss of marine sources of food is an important cause of increasing prices of grain. The demand for grain now rises directly with population increases. Unfortunately, we are no longer able to easily expand the amount of land under grain cultivation. We need large and continual increases in the amount of water for irrigation if we are to expand crop production. Even more precarious areas, such as the Sahel Zone of Africa and including the Nile River Valley, cannot increase their reliably irrigated areas unless they build enormous reservoirs such as the one at Aswan, which can store enough water to overcome several dry years. The days of easy increases in seafood harvests or crop production are past.

The loss of marine and freshwater shellfish, fish, and whale populations, which has dramatically sharpened at the end of the 20th Century, has moved humanity into a new and dangerous phase of food shortage. Unless we quickly learn to increase our reliable supply of water for irrigated crops while also restoring aquatic and marine habitats, we will see a tightening global food shortage in the very near future. Poorer countries will suffer famine. The grim stories of Ethiopia, Somalia, and North Korea will become commonplace.

2.3.3 Need for New Dams and Water Systems

To avoid this crisis, we will have to build more dams and new water systems to protect our water resources. The new dams will be needed for irrigation and water supply, including some very large dams for storage over long drought periods. We will also have to rehabilitate, modify, and perhaps remove many of the older dams on the east coast of the United States that are no longer productive and merely obstruct and prevent necessary migration of important fish. To design and build these new dams, canals, and related water projects, we need water planners and dam engineers who know how to avoid the mistakes of their predecessors. Changes in engineering education are required, and new approaches to river basin management are needed.

The need for a continuing emphasis on dam building makes ecologists and sociologists uneasy because of the damage that many dams have caused in the past. Persistent faults in some of the big dams have included the brutal displacement of rural, indigenous peoples for the benefit of urban populations and industrial interests, and the destruction of ancient and complex hydrologic and ecologic systems. Increased urban water supplies have been followed by contaminated rivers and damaged harbors because of the increased sewage flows. To avoid repetition of these unpleasant experiences, we need a new approach to water planning and dam building.

There is a justifiably jaded skepticism to appeals for new dams, water supplies, or sewage disposal systems. Many projects have been poorly planned and inadequately sized, and the needs for many projects have been evaluated through notoriously biased economic analyses. As a result, government funding for such projects has now been reduced enormously, despite steady increases in the need for such facilities. Clearly, the old ways are not working. A new way is needed.

Perhaps there is a chance in the Americas to combine the best of the technological base of the *immigrant* Americans with the sustainable principles of indigenous cultures of the *original* Americans, to develop a better living style for the *new* Americans of the Third Millennium. This new approach should be based on protection of our natural resources for optimum utilization, using principles of sustainable management and harvesting. It requires entirely new principles for water and environmental planners.

REFERENCES

Abramovitz, Janet N., 1996. Imperiled Waters, Impoverished Future: The Decline of Fresh-water Ecosystems. Worldwatch Institute Paper 128, Washington, D.C.

Bain, Mark B. and Boltz, Jeffery M., 1989. Regulated Streamflow and Warmwater Stream Fish, A General Hypothesis and Research Agenda. U.S. Fish and Wildlife Service, U.S. Department of the Interior. Biological Report 89(18).

Brown, Lester R., 1996. *State of the World. A Worldwatch Institute Report on Progress Toward a Sustainable Society.* W.W. Norton, New York.

Gore, Al, Senator, 1992. *Earth in the Balance, Ecology and the Human Spirit.* Houghton Mifflin, New York.

New York Times, 1970. Earth Day at 25: How Has Nature Fared? April 22, p. C-1.

U.S. Fish and Wildlife Service, 1995. Action Plan and Environmental Statement for 1995. The Silvio O. Conte National Fish and Wildlife Refuge in Massachusetts, U.S. Department of the Interior.

Dams, Canals, and River Basins

In Section II we review specific instances of ecologically important water projects along the New England coast, then from the shores of Hudson Bay in Canada, across the breadth of the United States, to the Caribbean Sea, and finally down to Mar de la Plata in Argentina. Numerous case studies are cited to illustrate important aspects in the historical development of American water resources. In addition to the ecological impacts of dams and canals, considerable attention is given to parallel deterioration in the water quality of lakes, rivers, and harbors.

The myriad dams of the New England coastal waters are over 100 years old, continuing in operation only through the "Grace of God and large safety factors," according to the late G. Williams of the Massachusetts Institute of Technology. But few of them serve their original purposes; most of them just damage the ecology of the rivers.

The history of the water and sewage systems for Boston Harbor is expounded in detail because they are the oldest water systems in North America, and because they are approaching a financial crisis that will need careful examination and management if severe ecological damage is to be avoided.

The importance of large shipping canals in affecting migration of exotic aquatic animals is illustrated in an examination of the St. Lawrence Seaway between Canada and the United States.

Toxic and hazardous waste problems affecting canals, reservoirs, and harbors are illustrated with three examples from North America. The long and tortured histories of Love Canal in New York and Neponset Reservoir and New Bedford Harbor in Massachusetts are used to demonstrate the recent emasculation of government programs to clean up industrial wastes. The striking influence of industrial and commercial corporations in modifying regulations for Superfund proceedings is portrayed.

Water resource planning is also analyzed for the tropical Americas. In the islands of the Caribbean Sea, dozens of hydroelectric dams were built after World War II, including 27 dams in the mountains of Puerto Rico. Fifty years later, the planning assumptions of the dam engineers have largely been negated by population growth, increased energy requirements, and the unyielding limit on available fresh water. The impact of these erroneous assumptions is described, including the recent and severe water shortages in Puerto Rico.

23

On the many rivers of Brazil, Uruguay, Paraguay, and Argentina, the largest hydroelectric dams in the world have recently been constructed. Many planning and engineering mistakes made in the U.S. in the past century are being repeated in South America now.

For some of the problems found in these case studies from both North and South America, specific solutions were derived, based on a broad analysis of historic and ecologic patterns. The most important is a proposal to prevent repetition of another ecologic disaster in Boston Harbor.

An institutional solution is proposed for abandoned dams in rivers of New England, focusing on restoration of habitats for migratory fish. The economic losses in an industrially contaminated reservoir in eastern Massachusetts are precisely calculated from property tax records over a 12-year period. This recent information, supplemented by a similar study of lakes in Maine, gives water planners and enforcement agencies more precise data to assess the economic impact of water pollution. The data should be useful to set fines for polluters, or to justify expensive remedial action.

Major faults in the process of water planning and dam design in the western U.S. are summarized for the benefit of dam builders throughout the Americas. The specific examples cited in Section II, of mistakes and local solutions across the broad range of conditions throughout the Western Hemisphere, provide a body of information from which general solutions are derived in Section III.

CHAPTER **3**

New England

3.1 HISTORICAL INTRODUCTION

We are the *Wabanakis* — Children of the Dawn Country, People of the East. Five tribes made up our nation — *Passamaquoddy, Penobscot, Micmac, Maliseet,* and a tribe now gone that lived on the Kennebec River.... Long have the white men been among us. Yet some of us still remember the time when our lives were spent in hunting and fishing, and our villages were of wigwams instead of houses.

In the olden time our garments were of moose skin and fur...we fished with a bait of stone, well greased with moose tallow, on a line of moose sinew. Our lives were simple and glad, and our marriages were happy. Man and woman made their vow to the Great Spirit. In our old religion we believed that the Great Spirit who made all things is in everything, and that with every breath of air we drew in the life of the Great Spirit. — ***History of the Wabanakis,*** a people of Algonkian stock from the Atlantic Coast to the South of the St. Lawrence River, as told by Bedagi in the 19th Century. (From *The Indian's Book,* 1987, edited by Natalie Curtis, Bonanza Books, New York.)

New England and other coastal cities of North America have suffered a great sea change in their economies at the close of the 20th Century. Although universities and technology-based industries have given the economy some degree of resiliency, their influence could not overcome the losses of fishing, whaling, and farming, which had always sustained the original indigenous peoples of the area, and which were quickly adopted by the immigrant population arriving from Britain and Europe in the 17th Century.

For the most part, these activities disappeared because of population growth and fundamental errors in the management of regional natural resources, especially the lakes, rivers, and oceans with their aquatic and marine life.

New England dams and rivers are not very large; the reason for describing them in detail here is that they have been evaluated more thoroughly and over longer periods of time than most river systems in the Americas. Over the last two decades, enormous and almost continuous water quality surveys have been conducted by government agencies, using the most current techniques and analytical methods.

Based on the extensive field data, intricate computer simulations of almost every one of the rivers in Massachusetts were developed. The models were used as operational regulatory tools to set allowable loadings for discharges and to set limits for discharge permits. The models incorporate hydraulic, chemical, and biological components, and are valuable representations of the aquatic ecology. Furthermore, they are documented in great detail in river reports issued by environmental agencies since the early 1970s. The locations of these invaluable and historic documents are listed in the appendices. Unfortunately, all of these documents point to serious problems in our attempts to manage a sustainable environment.

Dam builders, sanitary engineers, and environmental planners were principal actors in this mismanagement of coastal New England resources during the past three centuries. However, they received a great deal of help from political leaders, rapacious industries, and misguided whaling and fishing interests. Together with the swelling regional population, these actors gave a dramatic performance in the New England Tragedy of the Commons.

This historical review of the tragedy in New England was developed to provide the remaining planners and engineers with insights into, and guidelines for, avoiding such tragedies in the future. More importantly, principles were established for rehabilitating these resources, and managing them in a sustainable fashion as a way of revitalizing the regional ecology and economy. The future requires that these planners and engineers drastically revise their conceptual bases and approaches; thus, this book may not be pleasant for some readers.

3.2 CLIMATE AND HYDROLOGY

It is important to understand the climate and hydrology of this region, not only to comprehend the seasonal ecology of the rivers and bays, but also because some fundamental aspects of these seasonal changes have been neglected by dam, water, and environmental agencies in the past. In order to correct these omissions, it is helpful to examine the impact of hurricanes on the forests, rivers, and fish, and the impact of winter ice cover on toxic ecology in small reservoirs.

3.2.1 Summer Hurricanes

The history of hurricanes in New England usually begins with the 1938 hurricane that drove straight up Rhode Island Sound into Narragansett Bay (Figure 3.1). Since then, there were two major hurricanes in 1955, as well as several smaller hurricanes in the last decade. The Atlantic hurricane season of 1991 was the most prolific ever, but only Hurricane Bob reached New England on August 19th. It drove ashore slightly east of Narragansett Bay, doing considerable damage along the Rhode Island and Massachusetts shores near the state line.

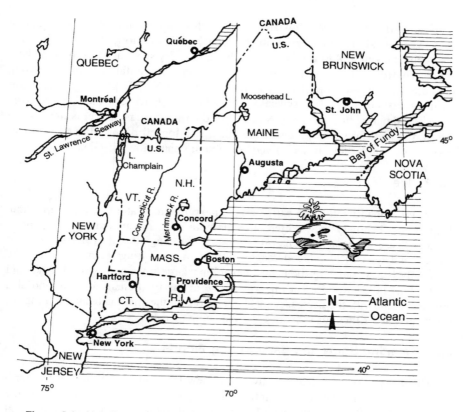

Figure 3.1 New England states and major bodies of water. (Map by Paul Metcalf.)

3.2.2 Autumn Droughts

The driest season of the year in New England is the period from August through October when there is usually little or no rainfall, and the winter storms have not yet started. This dry period was the reason for building most of the industrial dams along the New England coast: to store spring floods for use in the autumn. Statistical calculation of design droughts for water quality predictions are usually based on analysis of the driest weeks of record for each year, making an estimate for this low 7-day flow that might be expected every 10 years. The required degree of treatment of sewage, or the allowable load of organic matter, nutrients, or toxic substances is then calculated with the assumption that only this amount of flow will be available to dilute the sewage during periodic droughts.

This characteristic rainfall pattern is also one of the reasons why agriculture has generally failed in New England, except for lowlands along the river beds where water can be stored easily in low reservoirs. The rains often fail early in the crop year, many times by June or July. Thus, upland crops are stunted by lack of water during the growth phase, and then often wilt and die if the drought continues into August when they normally mature.

Cranberries that grow in lowlands are one of the few profitable crops remaining in New England. They can grow in the sandy, acid soil along the rivers if small reservoirs are kept full to get them through the long dry spell, and also protect them from early frosts. But the generally thin topsoil, adverse rainfall pattern, and strong pressure for real estate development have driven most New England farmers out of business. Except for a few specialty crops like cranberries, apples, sweet corn, squash, and pumpkins, the crusty New England farmers can no longer compete with the Midwestern farmers who enjoy the thick, rich topsoil of the Mississippi River floodplains, and the ideal rainfall and temperature pattern of the Great Plains.

3.2.3 Ice

New England winters are extremely variable in temperatures and precipitation. January and February thaws often alternate with blizzards and early spring snow-storms. These irregular patterns often produce flooding situations in the midst of winter, and do considerable damage to trees in the spring when snow can be extremely heavy, breaking branches and felling large trees across the brooks and rivers. This fallen wood is an important base of the food chain for fish, and also provides them with the necessary local conditions in the small tributaries of rapids, backwater, and riffles for spawning and hatching of fish.

3.2.4 Melting Snow and Early Spring Floods

The design flood for many spillways in New England is now the spring storm of 1936, which combined heavy rainfall with ice jams on the rivers, causing flooding far in excess of that expected based on the rainfall. However, most environmental studies during the Green Movement were designed only to detect conditions during the autumn droughts. Other seasons were neglected. Then, in the 1990s, one began to see the importance of winter conditions. Studies on Neponset Reservoir in Massachusetts showed the special importance of winter conditions and hurricanes in the ecology of industrial dams and contamination of their reservoirs.

There are many beautiful features of New England ecology, but at the end of the 20th Century, the small industrial rivers were not among them. New England forests are magnificent, especially in the autumn when foliage riots in colorful explosions. The craggy coastline and its superb islands are another. But these poor, desecrated rivers have been trashed for over a century. One can easily recognize when one is approaching an industrial river in New England. Wherever there are abandoned mills, shabby housing, junkyards, scrap heaps, razed brick buildings, and trash, there are the rivers of New England. At the end of the 20th Century, the rivers became so foul, so putrid, and so desecrated that mill workers normally went away from the river to eat their lunches. They could not stand the greasy brown slime and the odor of rotting algae.

New England is home to innumerable small dams on innumerable tributaries of once ephemeral streams. The purposes of the dams are long forgotten, but the dams continue to restrict the rivers and their fish, and impede the life cycles of migratory species. If we are to restore these basic fisheries that once sustained the Algonkians, we *new* Americans who now inhabit the banks of these streams must become better stewards of these gifts.

3.3 MASSACHUSETTS BAY

As the centerpiece of the New England economy and culture, Massachusetts Bay epitomizes the history of the northeast, now in serious ecologic decline. The mismanagement of natural resources of Massachusetts Bay and Boston Harbor received national attention in the 1988 U.S. presidential campaign, and will continue to symbolize an entire era of unsustainable action. Mistakes by environmental planners and engineers have been set in concrete in Boston Harbor, and will be visible for years to come.

Prior to the arrival of the British and European immigrants to Boston in the 17th Century, the pristine waters of Massachusetts Bay supported a small population of indigenous peoples with abundant fish, shellfish, crustacea, and marine mammals.

Massachusetts Bay receives flow from the Merrimack River in northern Massachusetts, the combined flows of several rivers entering Boston Harbor, as well as that of a few small coastal rivers just north of Cape Cod (Figure 3.2). Massachusetts Bay also has a unique sea-level connection with Buzzards Bay through the Cape Cod Canal.

A prominent feature of Massachusetts Bay, in addition to Boston Harbor and Plymouth Rock where the Puritans first landed, is Stellwagen Bank, former feeding ground for migratory populations of whales, and still a tourist attraction for spring and autumn whale-watchers.

The principal rivers entering Boston Harbor are the Neponset, Charles, and Mystic Rivers, as well as the small Weymouth Fore, Weymouth Back, and Hingham Rivers. A great deal of the story of the development of New England in the last three centuries can be found in the dams and mills along the Neponset and Charles Rivers; so we will begin our tour there.

3.3.1 Neponset River

The headwaters of the Neponset River originate in Neponset Reservoir in Foxboro, south of Boston. Most of this small basin is within the metropolitan Boston area, and the river ends in Dorchester Bay, one of the original settlements of the Massachusetts Bay Colony, formed soon after their landing at Plymouth Rock (Figure 3.3).

From Foxboro to Dorchester Bay, the Neponset River falls 270 feet over a distance of 30 miles, draining 117 square miles of heavily populated land, including a large variety of light industries. The river is impounded by 12 dams and passes through several privately owned reservoirs (Figure 3.4). Dams and reservoirs on the river date from the earliest years of colonial settlement. Most of the existing dams on the river are descendants of a long series of primitive dams constructed since 1634.

The first canal constructed in the Western Hemisphere in 1640 is a tributary of the Neponset River in Hyde Park. This canal was dug from the Charles River in

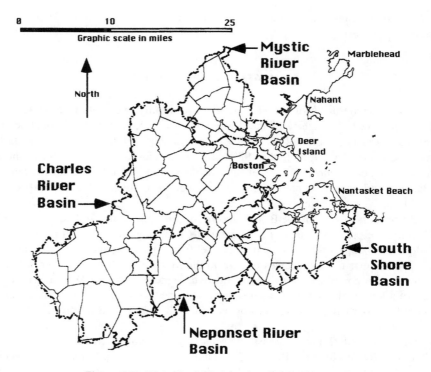

Figure 3.2 Watersheds draining into Boston Harbor.

Dedham along the original path of Mother Brook to supply additional flow to the mills in Milton at the Neponset River estuary. At the headwaters of the river, Neponset Reservoir in Foxboro was constructed in 1845 by the Neponset Reservoir Company, representing the several mill and dam owners along the river.

3.3.1.1 Neponset Reservoir in Foxboro

The most important source of ecological damage to the Neponset River at the end of the 20th Century was related to industrial wastes from the Foxboro Company, from paper mills, and from other facilities of the Neponset Reservoir Company. These affected the Neponset Reservoir in Foxboro at the very beginning of the river, and several smaller reservoirs on the Neponset River and its tributaries in Walpole.

Severe industrial contamination of Neponset Reservoir by the Foxboro Company occurred in the 1970s and 1980s, damaging the ecology of the reservoir and having serious consequences on the suitability of the entire river for fishing and swimming (Figure 3.5). Several local, state, and federal groups conducted water quality surveys to determine the nature of the contamination.

The monitoring of conditions in Neponset Reservoir yielded valuable information about heavy metal contamination of such industrial impoundments. Ironically, it was a group of volunteers who obtained much of this information, which had been missed by all of the professional scientists and engineers who conducted the multitude

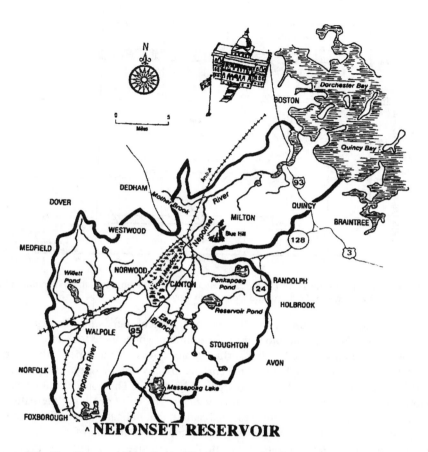

Figure 3.3 The Neponset River Basin. The location of Neponset Reservoir is shown in Foxboro, headwaters of the Neponset River.

of federal and state surveys. The volunteers were the senior chemistry students from Foxboro High School and their teacher.

Student Insights — As a supplement to the available information on the reservoir, the Foxboro students and their advisors decided to collect monthly measurements of heavy metals in the reservoir. Starting in February 1991, they continued through December 1991. The concentrations of metals in the surface waters were fairly high, even though the Foxboro Company had terminated its discharge in 1989. Apparently, the residual metals in the sediments were redissolving or being suspended in the overlying waters. During the spring and early summer, concentrations of copper and phosphates were extremely high in the surface waters, but cadmium and other heavy metals were not detected.

On August 19, after the long hot summer of 1991, Hurricane Bob passed near Neponset Reservoir with winds of 50 miles per hour gusting to 85 miles per hour, and a total rainfall of 4.5 inches in 24 hours. This produced a dramatic increase in

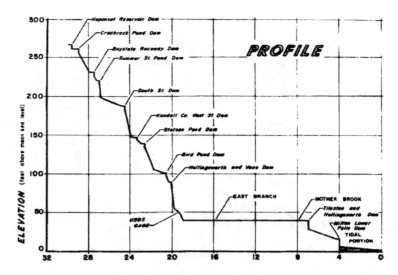

Figure 3.4   The Neponset River profile, showing location of 12 dams on the mainstem of the river between Foxboro and Dorchester Bay.

the concentrations of cadmium in the reservoir waters, persisting throughout the next 4 months. The concentration of total cadmium, which includes suspended and dissolved cadmium detected in the surface waters, was 18 parts per billion (ppb) during September, much higher than the chronic toxicity criteria of the U.S. Environmental Protection Agency (EPA) of 0.4 ppb (Figure 3.6). The concentration rose in October, reaching a peak of 27 ppb in November, gradually declining thereafter.

There were two aspects of this seasonal pattern that indicate the need for a major change in current methods of assessing cadmium contamination of reservoirs, and perhaps for other heavy metals. First, the past practice of summer surveys for measuring heavy metals in rivers and reservoir waters has probably missed the true toxic impacts of cadmium and other metals. Unless the surveys were conducted after large summer or autumn storms such as hurricanes, the toxic metals such as cadmium would not be in the surface waters, but would be immobilized in the sediments. Future summer surveys should be planned to coincide with large storm events, such as hurricanes, in order to measure the ecological disruption and temporary toxic conditions.

Second, additional surveys should be conducted in the cold months of late autumn and early winter to measure the maximum concentrations of cadmium and other metals, rather than doing them only in the comfortable warm months of spring and summer. The traditions of summer surveys were strongly rooted in convenience, availability of additional summer help from college students, and the pleasure of working in warm weather. Although this period does coincide with many of the severe episodes of algae population explosions, oxygen depletion, and low river flows, supplemental winter surveys will be needed in the future for industrially contaminated rivers and reservoirs to monitor impacts of heavy metals.

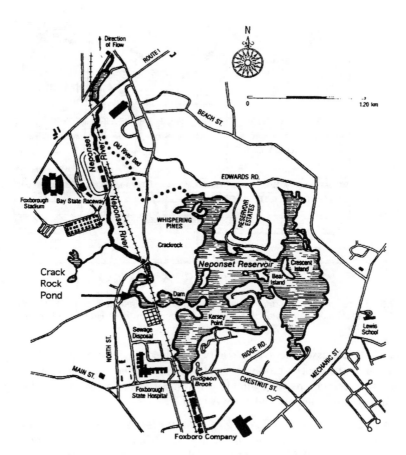

Figure 3.5 Neponset Reservoir in Foxboro.

Confirmation of Foxboro Student Findings — In 1994, unrelated to the discoveries on Neponset Reservoir by the Foxboro students, but directly confirming their validity, the EPA issued tentative sediment quality criteria for heavy metal toxicity that made the same conclusions about the need for seasonal surveys, and the benign impact of cadmium and other metals in sediments during the spring and summer seasons.

In 1994 and 1995, the federal and state agencies began supplementing their intensive summer surveys with winter surveys, especially aimed at gathering data on the toxicity of sediments. Their biological assessment of sediment toxicity in Neponset Reservoir indicated that the concentration of cadmium in the surface waters in November 1994 was at severely toxic values. Only 2% of small amphipod organisms survived exposure to the sediments. Similar tests the following summer indicated no toxicity, despite high concentrations of heavy metals in the sediments.

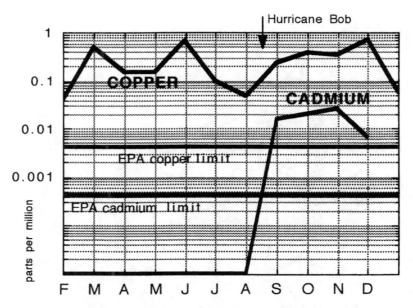

Figure 3.6 Heavy metals in surface waters of Neponset Reservoir in 1991. Graph shows seasonal variations and impact on total cadmium from Hurricane Bob on August 19th.

The biological assessments of toxicity coincided precisely with the findings of the students on the concentrations of total cadmium in the water column, in which the highest concentrations were found in November, but none was found during the quiet summer period in July and early August.

Sulfides in Sediments and Toxicity of Heavy Metals — The current theory is that sulfides in the sediments bind the heavy metals under certain conditions, such as those occurring on the bottom of these lakes in spring and summer. Suspension in the surface waters due to strong winds or cold weather conditions release the sulfide bonds, freeing the metals to exert toxic effects on small aquatic organisms.

It would be a mistake to assume that the warm-weather binding of the heavy metals is protecting the ecology of these industrial reservoirs. During windstorms when the sediments of the reservoirs are brought into suspension by waves and currents, the toxicity of the metal can suddenly be exerted. If this occurs in the spring, immature stages of fish and small aquatic organisms will be severely affected. Under the ice in Neponset Reservoir during the winter, it was also common to see large numbers of dead fish appearing suddenly, perhaps because of the toxic manifestation of these heavy metals stored in the sediments. The death of smaller organisms at the base of the food chain is not seen directly, but shows in the distorted assemblages of fish and other organisms.

Recent Changes in Industrial Discharges — Although the companies that discharged industrial pollutants to the Neponset River were gradually forced to stop in the 1980s, most of them simply switched their discharges to the Massachusetts Water Resources Agency (MWRA) sewers, thus transferring the wastes further downstream to Boston Harbor, but not really eliminating the problem. The Foxboro Company redirected its wastewater discharges to the Mansfield sewage facility, thus merely changing its discharge to the Three Mile River at a point where the flow was larger than that of the Neponset River in Foxboro. However, this company also discontinued its plating processes because of the toxicity of cadmium.

Furthermore, the increasing sewer and water charges of the MWRA due to its large construction program of the 1990s caused one of the Neponset companies to renew its industrial discharge to the river in East Walpole, opening the scene for a repetition of the problems of the previous century when the river was a foul sewer. The past century of experience with industries on this river clearly demonstrated that they would ruin the river unless they were strongly regulated by environmental agencies. As environmental agencies were losing their effect in the waning years of the Green Movement, industries would once again be able to increase profits by dumping wastes in the Neponset River.

The historical disregard of industries along the Neponset River for its welfare was physically apparent in the trash, discarded machinery, and crumbling buildings along its banks in East Walpole and Norwood (Figures 3.7 and 3.8). While this attitude continues, it will be hard to reclaim the river by current governmental mechanisms.

There is an important and little appreciated sociological problem encountered by small and weak regulatory agencies trying to regulate large and wealthy industries. In the long series of meetings needed to implement complex environmental regulations, the representatives from the industry assert their social dominance. The regulators gradually lose their environmental orientation, adopting instead the concerns and interests of the industries they are supposed to regulate. The same is true for large cities that can afford expensive lawyers and engineers as their representatives.

Conversely, the field scientists, who see the rivers and harbors every day and appreciate the magnitude of the industrial contamination, are seldom present at these meetings. Even without bribing government enforcers, the industries are able to overcome the intent of the regulations by social manipulation of the environmental agency representatives. Thus, initially fervent enforcement efforts of health and environmental agencies gradually deteriorate into largely ineffectual programs, especially regarding large industrial polluters. So ended the Green Revolution of the 20th Century.

As long as the regulatory measures of state and federal agencies were poorly based and weakly organized, the industries were disinterested in cleaning up their mess. When the paper mills of the Neponset Reservoir Company started discharging their wastes at the end of the 19th Century, they installed no treatment devices whatsoever, despite monitoring by the state Department of Health and strong complaints from local citizens.

Figure 3.7   Deteriorating spillway on Bird Company Dam on Neponset River in Walpole.

This behavior was repeated a century later by the Foxboro Company in 1994 when the Town of Foxboro asked the company to repair the ecological damage it had done to the Neponset Reservoir and the river downstream. The Foxboro Company ignored the request. Regulatory proceedings were initiated against them by the state Department of Environmental Protection in 1995 under hazardous waste regulations, but the company delayed action, thus avoiding the expense of correcting the problems it created.

Similar histories occurred in more toxic contamination episodes, such as Love Canal in New York and in New Bedford Harbor on Buzzards Bay of Massachusetts. It is clear that new arrangements are needed.

The Foxboro Company Reservoir Debacle — Neponset Reservoir was constructed by the Neponset Reservoir Company in 1845. Soon afterward, the Foxboro Company (not related to the Neponset Reservoir Company) was established by a local family of engineers, expanding to an international corporation by mid-century, manufacturing industrial equipment. For the first part of the 20th Century, Neponset Reservoir was an ideal fishing ground for all manner of fish, including prize bass (Figure 3.9). However, in 1970, this company established a metal plating plant with a discharge to a small brook flowing into Neponset Reservoir. The trouble actually began in 1970 with this industrial discharge, but serious problems did not surface until 1984, which is designated as "Year 1 of the Debacle."

The Foxboro Company debacle went on for 14 years and may continue for many more, especially if previous histories of hazardous industrial waste sites are any guide. For example, the procedures and law suits for cleaning up Love Canal lasted from 1977 to 1996 — 19 years without reaching complete resolution. Even worse,

Figure 3.8   Deteriorating mill buildings along channelized Neponset River in Walpole.

the process to restore New Bedford Harbor in Massachusetts began with closing the harbor to fishing in 1979 and demanding payment for the cleanup from the five principal polluters. Although a payment of $109 million from these industries was finally received by the EPA in 1992, the cleanup did not begin until 1995, over 16 years after the problem was identified. Complete restoration of New Bedford Harbor could easily take 25 years. A final example is the contamination of the Housatonic River and land around Pittsfield, Massachusetts, with toxic and carcinogenic polychlorinated biphenyls (PCBs) by the General Electric Company. The contamination was discovered 20 years ago and is now known to be widespread, even down into Connecticut. But only one Superfund site has been designated, leaving about 100 miles of river contaminated with this industrial toxin. The industries have immense powers of resistance.

From the history of the Foxboro Company debacle, it is clear that if remedial actions are ever taken by the Foxboro Company, they will not start until more than 14 years after the problem was clearly evident. At least half a dozen large-scale studies were conducted on this small 300-acre reservoir, and endless meetings were held to move the restoration process forward. At the end of all this, there is little guarantee that the real problems will be corrected. Furthermore, the ecological damage of the ensuing 14 years can never be corrected. There must be a better way.

Figure 3.9   Bass used to enjoy the diverse and stable ecology of Neponset Reservoir. Before the Foxboro Company debacle began, Neponset Reservoir was regionally celebrated for good fishing. Prize-winning bass were often caught by anglers, who also enjoyed the quiet beauty of the setting. (Drawing by Fran Eisemann.)

A principal cause of the incredible delays is that state and federal hazardous waste procedures are not designed for complex or serious situations. One wonders if the rules and regulations were specifically influenced by the lobbyists for the industries so that they could use them to delay action and expenditures for any serious problems.

If the problem is small and clear-cut, such as an oil spill from a vehicular accident, and the responsibility rests with only one firm, then the procedure might be fairly efficient. But even for such a simple case, it would be easy to delay action for at least 3 years. Anything with multiple responsible parties and complex ecology is open to decades of stalling by the responsible groups. The rules and regulations need to be completely rewritten to make it possible to remedy the hazards in a short time.

There is an urgent need to find a better way to deal with these problems, especially regarding contaminated reservoir, lake, and river sediments that have seldom been regulated in the past. The industrial dams of New England and their contaminated reservoirs are major obstacles to restoration of natural fisheries along the coast, and constitute a long-term threat to marine life as well.

On the Neponset River, the necessary conditions are present for a new strategy for restoration. Many of the old mills are abandoned, their grounds strewn with trash and rubble. They have very little real estate value because, not only are the buildings ruined,

---

Chronology of Foxboro Company Debacle, 1984–1997

Year 1. Company discharges phosphoric acid for cleaning metal plating plant. Complaints by shoreline residents of algae and foul odors.

Year 2. Local citizens' group formed in response to foul odors and unsightly algae mats covering reservoir throughout the summer. Group is called Neponset Reservoir Restoration Committee, and asks Foxboro Company to stop discharges and clean up reservoir. Company reduces amount of phosphates in discharge but continues plating operations.

Year 3. After insistent complaints by Restoration Committee, a thorough investigation is conducted by state and federal environmental agencies.

Year 4. Issuance of revised discharge permit requirements causes Foxboro Company to curtail principal offending discharges. Severe problems persist in reservoir, however.

Year 6. Town hires environmental firm to conduct diagnostic and feasibility study of reservoir problems. Schools, civic groups, and local firms, including Foxboro Company, join in study. Foxboro Company reduces discharges further.

Year 9. Town requests Foxboro Company to restore reservoir. Company takes no action.

Year 10. Instead of restoring reservoir, Foxboro Company hires its own consultant, who concludes that there is no problem. Report of first consultant rejected by town because of serious technical flaws. Company supports its consultant and contests report of town environmentalist. Town hires independent third firm to evaluate conflicting reports. Third firm concludes Foxboro Company report was in error.

Year 10. Federal and state agencies conduct winter survey of reservoir and downstream impoundments and find clear toxicity of sediments to amphipods and other bottom-dwelling organisms.

Year 11. After heavy lobbying by town government and citizen groups, state Department of Environmental Protection issues Notice of Responsibility under hazardous waste laws to Foxboro Company. Foxboro Company hires full-time environmental lawyer to handle problem and hires new consultants. Consultant conducts large study in August and September, duplicating several previous studies but missing toxicity of heavy metals. Despite advice from Technical Advisory Group of state and federal scientists, Foxboro Company consultant does not conduct survey in winter when metal toxicity would be evident.

Year 12. Phase I of voluntary study completed. Classification 1A proposed by Foxboro Company. Phase II initiated for next 2 years. Intent is to conduct complete risk assessment analysis and determine remedial measures.

Year 13. Storm drains under Foxboro Company buildings found to intermittently discharge toxic metals and industrial solvents into reservoir and aquifer.

Year 14. Phase II studies to be completed, perhaps. Then remediation activities should start, perhaps.

---

but the river is polluted. These abandoned mill sites should become the location of new treatment facilities for the removal and detoxification of their own reservoir sediments. They might also be used for instream treatment of the river in the summer when surges of toxic metals and nutrients are flushed down the river by storms.

Many of the old mills had sedimentation tanks or other simple treatment facilities for their industrial processes. Some of these might be modified directly for use in restoration of the river. Rather than imposing fines or lawsuits on the industries that ruined the river, it might be better to have them install treatment facilities on their properties, and use them to restore the river.

Loss of Real Estate Values Around Neponset Reservoir — Neponset Reservoir, at the head of the Neponset River, epitomizes the problems of small reservoirs in Massachusetts. Increasingly offensive conditions in Neponset Reservoir during the 1980s reduced the attractiveness of the reservoir for recreation and esthetic enjoyment and reduced property values of residential lots around its shores, as well as causing damage to the fisheries. To quantify some of these impacts, an estimate was made of the financial loss suffered by landowners along the reservoir shore. This was accomplished by comparing changes in real estate values around Neponset Reservoir from 1983 to 1995 with changes around Lake Massapoag in Sharon, a similar and nearby lake that did not suffer from contamination.

The property values in 1983 and 1995 were obtained from town assessment records, for all properties that abutted the shores of the two water bodies. The properties compared in the study were further restricted to those properties that were not further divided into smaller parcels, and that did not have new houses built on them during the study period. Each property was listed for its full value assessment, as well as its area in square feet and its lakeshore frontage in feet.

The lakes had similar ecology, including a similar history of real estate and recreational development until 1970 when the Foxboro Company began to discharge wastes into Neponset Reservoir. Since that time, Lake Massapoag has continued to be an important recreational asset to Sharon, while Neponset Reservoir has turned into an unsightly and obnoxious nuisance in the summer.

The mean values for residential properties around Neponset Reservoir were lower than the mean values for residential properties in the rest of Foxboro, while the mean values for residential properties around Lake Massapoag were higher than the mean values for all residential properties in Sharon. Furthermore, during the 12-year study period from 1983 to 1995, the lakeshore values in Foxboro did not increase as fast as the values of residential properties in the rest of Foxboro, while the lakeshore values in Sharon increased at a rate faster than the values in the rest of Sharon.

In 1983, the mean value for the study properties around Neponset Reservoir in Foxboro was $51,781, slightly lower than the mean value for town properties of $56,693. In 1983 in Foxboro, the lake/town ratio for the mean values of residential lots was thus 0.91. By 1995, the mean value for the study properties had increased to only $126,711, a smaller increase than that of town properties that had risen steeply to $181,889. Thus, for 1995, the lake/town ratio had decreased to 0.70, a relative decline in the value of lakeshore properties in Foxboro.

Although all property in southeastern Massachusetts had been rising in value over the previous several decades, this comparison of the specific real estate values in the towns between 1983 and 1995 showed that residential lots around Neponset Reservoir in Foxboro were worth less than average residential lots in Foxboro as a whole. The annual rate of increase in value of the property around Neponset Reservoir was also low, about one half of the 6.8% increase for all residential property in Foxboro.

In contrast, property around the relatively clean Lake Massapoag in Sharon was worth more than ordinary residential property in Sharon, and was rising in value at a rate faster than that of ordinary property. In 1983, the ratio of property values for residential lots around the lake, compared to residential lots in the entire town, was 1.09, rising rapidly by 1995 to a lake/town ratio of 1.17 (Table 3.1). These ratios were used in projecting property values for Neponset Reservoir under projected clean water conditions of Lake Massapoag.

The mean area of the 83 study lots around Lake Massapoag was 0.7 acres, slightly smaller than the mean area of 0.8 acres for the 48 study lots around Neponset Reservoir. Nonetheless, the value of the lots around Lake Massapoag was higher than that of the study lots around Neponset Reservoir. The northern shore of Lake Massapoag was a public beach, and the southern shore was a public park with a community center and launching facilities for sailboats (Figure 3.10). The lake was the central focus for the annual Fourth of July celebration in Sharon, including the community fireworks display.

Although 94 lots had shorelines on Neponset Reservoir in 1995, a house-to-house survey in 1990 indicated that the number of residential properties immediately affected by conditions in the reservoir was 158, due to several small neighborhoods tightly clustered along the shoreline. In 1983, the number of houses in this same group was about 110, based on dates of house construction obtained from the 1990 survey. It was thus estimated that the number of houses directly affected in 1995 was 160.

The total economic impact of the contamination on property values around Neponset Reservoir was calculated by estimating the additional value the property would have if the clean water conditions of Lake Massapoag existed in Neponset Reservoir, and thus the corresponding ratios of lake/town values were the same as those in Sharon (Table 3.1).

The lake/town ratios of property values in Sharon for Lake Massapoag indicated what could be expected for the Town of Foxboro if Neponset Reservoir had no contaminants. This was designated the "clean lake ratio." This value was then multiplied by the mean value from all residential lots in Foxboro to obtain the mean value for a lakeshore lot around a clean Neponset Reservoir without contamination. This value was then multiplied by the total number of properties around Neponset Reservoir to obtain their total value under clean lake conditions (Table 3.1). The difference between the total value for clean lake conditions and the total value for actual conditions thus indicated the effect of the contamination. These data were then projected into the future, based on the trends from 1983 to 1995.

These calculations for the value of real estate close to the reservoir underestimated the true losses in four ways. First the polluted reservoir has a smaller but more widespread effect in depressing real estate values throughout Foxboro. Second, the recreational and esthetic loss to the entire town represents a significant loss. Third, the study properties were limited to those that were already occupied in 1983, excluding the large number of lots and homes developed since then, around the reservoir. Many of these newer homes are much more expensive than the older homes; thus, the impact on their value is above the average impact calculated in this study. Fourth, no accounting was included of the many years of foul conditions in the reservoir during which the fish, turtles, birds, and other animals suffered from

Figure 3.10   Lake Massapoag in Sharon. The northern shore at the upper edge of the photo is a community beach. Interstate 95 passes the lake on the west, and the Great Cedar Swamp to the south was allegedly a hideout for Wampanoag warriors during King Phillips War.

the contamination. The damage that occurred to the reservoir ecology during these past years cannot be recovered just because the toxic materials are removed. The sum of these four effects might easily double the economic losses estimated in the following projections.

The trend of losses based on these data was projected into the future to indicate the expected values if no remedial action is taken. Because the ratio of lake/town

Table 3.1 **Projections for Property Values Around Neponset Reservoir if Clean Lake Conditions Existed Similar to Those in Lake Massapoag, 1983–2000**

Item for analysis	1983	1995	1996	1997	2000
PROJECTED CLEAN WATER CONDITIONS AROUND NEPONSET RESERVOIR					
Sharon Ratio for clean water conditions; lake/town value	1.09	1.17			
Mean value for residential lot in Foxboro, in dollars	$56,693	$181,889			
Number of lots around Neponset Reservoir	110	160	160	160	160
Actual total value of residential property around Neponset Reservoir with contamination, in millions of dollars	$5.7	$20.3			
Total estimated value of residential property around Neponset Reservoir with clean water conditions of Lake Massapoag, in millions of dollars	$6.8	$34.0			
Difference in property values around Neponset Reservoir between clean and dirty water conditions, in millions of dollars	$1.1	$13.7	$14.7	$15.8	$18.9

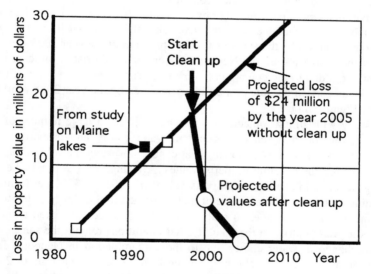

Figure 3.11 Projected decreases in real estate losses for Neponset Reservoir. These calculations are based on the assumption that the reservoir will be cleaned up by the year 2000 and will regain its real estate value by 2005.

property values was increasing rapidly for the clean water conditions in Sharon, this difference escalates every year at a rate of about $1 million.

From this simple extrapolation, the projected difference in real estate value around Neponset Reservoir would be $15.8 million by 1997, and $18.9 million by the year 2000 (Figure 3.11). In terms of losses to individual lot owners, the average loss in property value per owner in 1997 for 160 owners was similarly projected to be about $100,000.

In 1997 a question remained about the portion of the contamination in the reservoir that was due to discharges from the Foxboro Company. Following procedures in the Massachusetts Contingency Plan for Waste-site Cleanup, a year-long intensive study was scheduled to start in 1998. The conditions of this study were prescribed under Phase II, in accordance with the Notice of Responsibility issued in May 1995 to the Foxboro Company. Analysis of the results from the Phase II study were expected to indicate the role of the Foxboro Company in contamination of the reservoir. If their portion of the contamination is significant, they will be required to develop a remediation plan, and to carry it out.

If the offending contaminants are removed, and Neponset Reservoir is returned to conditions similar to those in Lake Massapoag of Sharon, the economic losses to property owners around Neponset Reservoir would be reduced markedly. The theoretical impact of this remediation was compared with a "no-action" projection, assuming the remedial activities would take 3 years starting in 1998, and assuming that the clean lake ratio from Sharon could be reached slowly, by about the year 2005 (Figure 3.11).

Because considerable time would be required for the lake to recover from the trauma of the contamination and from the remedial activity, and perhaps some additional time would be required before Neponset Reservoir regained its reputation in the real estate market as an attractive lake, the lake/town ratio for Neponset Reservoir might not be the same as that in Sharon until the year 2005. By then, the value of the properties around Neponset Reservoir would rise significantly and the economic loss due to the contamination should drop to zero. However, if the remediation is not carried out, the property loss expected by the year 2005 would be $24 million (Figure 3.11).

A study with similar objectives was conducted for a large number of lakes in Maine. A relation between differences in water clarity and property value was obtained from this analysis, showing that changes in water clarity in a lake caused a large change in shoreline property values. A change of 1 meter in Secchi disk readings resulted in changes in property values from $11 to $200 per frontage foot, depending on the local innate price of real estate, the initial clarity of the water, and other factors. Thus, changes in real estate values can be calculated from data on changes in water clarity in the lake, if other conditions remain the same.

The change in water clarity due to the contamination in Neponset Reservoir resulted in a decreased Secchi disk reading from the expected value of 2.0 meters prior to the contamination in 1970, to the value of 0.4 meters observed in 1992, a decrease of 1.6 meters. Combining the data from the Auburn area of Maine, where prices most closely approximate those of Massachusetts, with the loss in clarity observed in Neponset Reservoir, yields an estimated loss of $200 × 1.6, or $320 per frontage foot. Neponset Reservoir has a total shoreline of about 42,000 lineal feet. For a reservoir with 42,000 frontage feet, this would cause an economic loss up to $13 million for 1992.

The estimate from lakes in Maine was slightly higher than the estimate of a $10 million loss for 1992 obtained from this study (Figure 3.11), confirming that many million dollars of property value were lost around Neponset Reservoir because of the effect of contamination on its water clarity and quality.

Table 3.2 History of Water Quality Surveys on Neponset River, 1875–1994

Year	Agency	Number of parameters measured	Number of stations sampled	Focus of sampling	Ref.
1875	State Board of Health	7	7	River	Jobin, 1993
1891		7	12	River	Jobin, 1993
1895		7	14	River	Jobin, 1993
1938	Federal Works Progress Administration			River	Jobin, 1993
1964	State Department of Public Health	13	16	River	Jobin, 1993
1973	Division of Water	13	17	River	DWPC, 1973
1975	Pollution Control	15	11	Polluters	DWPC, 1975
1978	(DWPC)	14	18	River	DWPC, 1978
1986		30	15	River	DWPC, 1986
1986		17	6	Polluters	DWPC, 1986
1986		31	6	Neponset Reservoir	DWPC, 1986
1991		21	21	River	Webber, 1992
1994	Office of Watershed Management	40	10	River and reservoirs	OWM, 1995

3.3.1.2 Mainstem of Neponset River

Much of the useful data on water quality in Massachusetts rivers derives from surveys conducted since the Great Depression, when river surveys were included in the agenda of the Works Progress Administration (Table 3.2). Under this program aimed at economic stimulation, in addition to hiring chemists and field personnel to do the river sampling, writers were commissioned to summarize the results of these surveys and record them for posterity. This unusual endeavor in the face of economic difficulty has given us a legacy of information on the original state of our rivers.

It is ironic — and disturbing — to realize that exactly the opposite strategy was used in the 1990s, in the face of a smaller economic recession. All of the people who conducted and analyzed the river surveys of the 1980s were laid off in the 1990s, supposedly to lower taxes and thus stimulate the economy. The rise and fall of the scientific organizations monitoring the rivers are clearly indicated by the recent history of studies on the Neponset River.

History of Neponset River Surveys since the 19th Century — The current office of Watershed Management of the Massachusetts Department of Environmental Protection conducted surveys on the Neponset River in 1964, 1973, 1975, 1978, 1986, 1991, and 1994 (Table 3.2). Some of the basic parameters measured in these surveys are summarized here and in other chapters. From a robust organization of about 100 people, however, this group has been cut to less than a dozen, and future reductions appear to be coming. This will be the end of a superb documentation of progress in pollution control. Furthermore, it probably means that the progress in restoring the river will also cease, because the field surveys provide the data needed for enforcement actions.

There are some early records of river surveys from 1875 when water quality was quite good, at least in the upper reaches of the river. Additional surveys were made during the following 20 years, probably due to the opening of several paper mills in Walpole and Norwood that severely polluted the river, as shown in the data from 1891 and 1895.

Industrial and Domestic Discharges — An important factor in determining water quality of the Neponset River was the number and type of industrial and domestic sewage discharges pouring contaminants into the river system.

Although the numbers during previous centuries can only be estimated, it is likely that most of the serious problems began after the middle of the 19th Century. By the late 1800s, railroads and technological progress favored the development of textile and paper industries that used large quantities of water.

Apparently, the first large polluting discharge started around 1890. The number of discharges probably peaked between 1940 and 1960, with about 20 separate outlets. State records indicated 18 discharges in 1964, dropping to 6 in 1986, but then rising to 7 in 1995.

Total Solids — One of the earliest parameters of contamination measured in rivers was total solids, also called total residue. In 1875, the concentration of total solids in the Neponset River between Walpole and Dorchester Bay was about 50 parts per million (ppm), probably mostly salts from natural sources (Figure 3.12). This can be taken as an indication of clean water conditions for local waters. The concentration remained constant throughout the 20 miles of surveyed river.

In contrast, a survey at the same points in 1895 showed 2000 ppm total solids as the river left Walpole, apparently due to the operation of mills at the present site of Bird Company and Hollingsworth and Vose Company in East Walpole. These industries began when adequate summer water supplies were secured by construction of the Neponset Reservoir in Foxboro, and smaller reservoirs in Walpole such as Bird Pond.

In 1978, almost a century later, the concentration of total solids was still high, about 300 ppm. Although the total solids dropped to 175 by 1986, the amount of total solids was still over three times as high as it had been in its clean condition a century earlier in 1875.

Suspended Solids — Suspended solids in a river are usually quite noticeable, even to the casual observer. Clean water has concentrations of suspended solids less than 5 ppm. When the concentration of suspended solids reaches 10 to 20 ppm they become visually noticeable and offensive. This parameter was not measured in the older surveys, but more recent data showed that the concentrations were extremely high in 1964, reaching peak values of almost 100 ppm as the river passed through East Walpole (Figure 3.13).

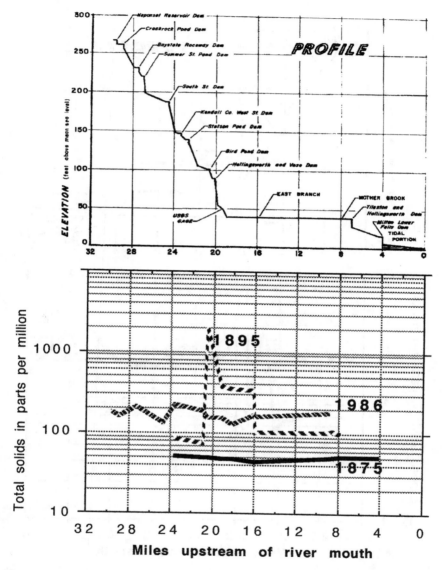

Figure 3.12   Over a century of variation in total solids in the Neponset River, 1875–1986.

Successive surveys showed a gradual decrease in suspended solids in the middle reaches of the river. However, in 1986, the concentrations in the upper reach showed a marked increase due to discharges from the Foxboro Company, the Foxboro State Hospital, and perhaps Foxboro Raceway. These recent measurements were far worse than conditions in 1964, and extended from Foxboro through Walpole (Figure 3.13).

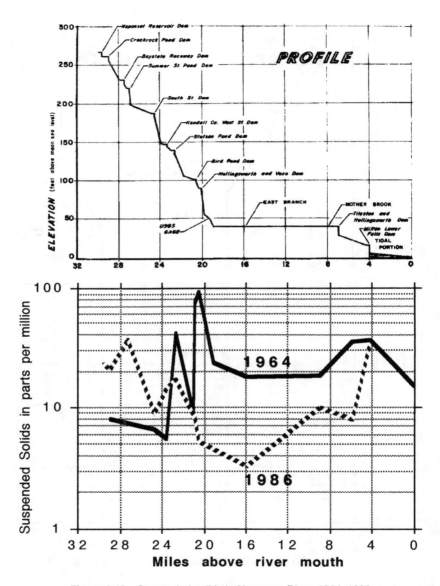

Figure 3.13 Suspended solids in Neponset River, 1964–1986.

By 1991, some decreases in suspended solids were noted. The mean for the entire river had dropped below 6 ppm. Improvement in Foxboro was clearly due to the closing of Foxboro Company's metal plating plant. This plant was closed in 1989 under pressure from local citizens because of severe degradation in the Neponset Reservoir, which had been receiving excessive phosphate nutrients and heavy metals from the Foxboro Company discharge.

Phosphorus — Phosphorus is a critical nutrient for aquatic algae and other plants, and when present in excess it causes objectionable growths of algae and weeds, especially in ponds and slow sections of rivers. Domestic sewage and industrial detergents are usually the principal sources of phosphorus in surface waters. Concentrations of phosphate salts above 0.05 ppm can cause objectionable algae blooms in late summer. In 1973, the concentration of phosphates exceeded this value throughout the river, especially upstream of Walpole Center (Figure 3.14). Concentrations in water leaving Foxboro exceeded 0.3 ppm.

Contrary to improvements noted in other parameters, there was a remarkable increase in phosphorus throughout the river by 1986, especially in the upper river where concentrations reached nearly 0.4 ppm. This was due primarily to sources in Foxboro.

By 1991, there was a noticeable decrease in phosphorus throughout the river, primarily due to elimination of industrial detergents from the Foxboro Company discharge in 1989.

Annual Loads of Industrial Solids — The concentration of solids in the river resulted primarily from the loads of solids discharged by industrial polluters, as well as from the solids coming from surface runoff. A useful characterization of the impact on the river is the total mass of solids coming annually into the river from discharges. This is calculated by multiplying the concentration of suspended solids in the discharge by the annual volume of the discharge.

The annual load of industrial solids discharged into the Neponset River in the Mid-20th Century probably exceeded 200 tons (Figure 3.15). Fortunately, this dropped to about 128 tons by 1973 and to 2.5 tons by 1986. Two and a half tons of suspended solids in the low summer flow of 10 cubic feet per second would result in a concentration of 8 ppm, about the values measured in 1986 (Figure 3.13). Thus, most of the suspended solids in the river were accounted for by industrial discharges. The major contributor to the 2.5 tons of solids in the Neponset River in 1986 was the Foxboro Company, the only remaining industrial discharge.

Diversion of Aquifer Flow to Harbor — By 1986, most of the industries along the river had connected their outlets to the MWRA sewerage system, diverting the wastes to Boston Harbor for partial treatment. Unfortunately, this also increased the amount of water that was being diverted out of the river. Many of the towns in the upper valley were pumping water from aquifers feeding the Neponset River. After this water was contaminated with human and industrial wastes, it was not returned to the river but diverted instead to the MWRA sewers and discharged to Boston Harbor.

As a result, the base flow of the river was gradually decreasing. In the dry months of late summer, the flow in the river comes primarily from the aquifer as rainfall runoff is near zero. When the aquifers are lowered by pumping, the flows into the riverbeds decrease. Low summer flows in the Neponset River estuary were a primary limitation on the natural smelt population in the 1990s.

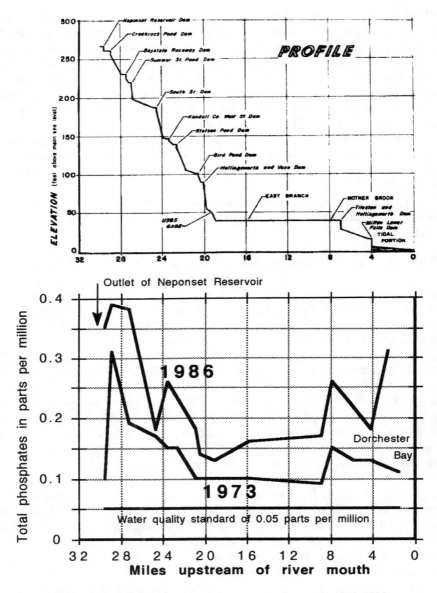

Figure 3.14 Total phosphorus in the Neponset Reservoir, 1973–1986.

Despite the diversion of industrial discharges away from the river, the upper 10 miles of the Neponset River in Foxboro and Walpole remained severely polluted at the close of the century. In 1995, the state announced that the upper mainstem of the Neponset River did not have the conditions necessary to support aquatic life, nor primary contact recreation, nor should fish from this reach be consumed, nor

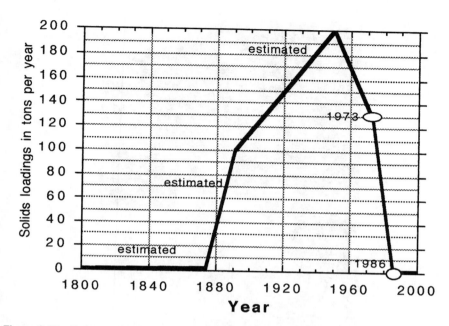

Figure 3.15   Estimated historical pattern of loads of solids discharged from industries into the Neponset River during the last two centuries.

was it esthetically satisfactory. This was a rather sad conclusion for one small river after 25 years of the Green Movement.

Both Neponset Reservoir and Crack Rock Pond slightly downstream were found to have severely toxic sediments and water quality, in addition to violating the primary standards for fishing and swimming quality. Biological assessment of toxicity of the sediments in Neponset Reservoir had shown that fewer than 2% of the test organisms survived 10 days of exposure to the reservoir sediments.

Restoration of Herring to Neponset River — One of the most obvious goals for restoration of the Neponset River is the reestablishment of conditions necessary for the smelt fisheries in the estuary, and the herring fisheries in the mainstem and the East Branch. Records from colonial times indicated flourishing herring runs up to Lake Massapoag in Sharon, the headwaters of the East Branch that flows through Canton. In other New England rivers, as many as 1 million herring made the annual run in 1996. This magnificent spring spectacle occurred only in rivers without dams in their lower reaches, such as the Nemasket and Taunton Rivers.

Given the persistence of these herring runs through three centuries of gross pollution, urbanization of watersheds, and over-fishing, there is good reason to believe that the herring could also run to Lake Massapoag in Sharon again. The productivity of the Neponset River system is high enough that these herring, and the shad, menhaden, and smelt that would probably also come back if the herring runs were restored, could restore an important addition to the local food supply as well as provide a great deal of local recreational activity in the spring seasons.

Figure 3.16 Spillway of Factory Pond Dam in Canton, first obstruction upstream of Fowl Meadow for migratory herring on Neponset River.

Primary obstacles to reopening the herring run are the two downstream dams on the lower river. Also, modifications must be made to the low dams on the East Branch and Sucker Brook coming from Lake Massapoag in Sharon, and the excessive heavy metals coming from the impoundments in Canton near the Plymouth Rubber factory, and from Neponset Reservoir on the mainstem of the river (Figure 3.16).

The heavy metals in the sediments of Factory Pond in Canton would be especially dangerous to migrating fish during spring storms, which churn up the pond bottom and cause the metals to release into the flow in toxic form. This spring-time hazard must be dealt with in permanent form because it represents a lethal hazard to the various life stages of the fish that migrate in the spring.

Similar obstacles exist in the mainstem, with toxic surges of heavy metals coming from Neponset Reservoir after storms, and also the physical barriers from several dams. Although the mainstem will require more remedial measures than the East Branch, the techniques needed will be similar. If progress is made in controlling the heavy metals and excessive eutrophication in Neponset Reservoir, the successful methods should also be used in Factory Pond of Canton.

If small hydroelectric power facilities can be installed in the lower dams near the estuary, using technology that allows fish passage in either direction, these installations can be evaluated for use further upstream where the reliable amount of flow is smaller. If the generation facilities are economically feasible, they could be used to generate funds for the other aspects of herring restoration.

Five million dollars in bonding were approved in an Open Spaces Bond Issue by the Massachusetts legislature in 1996 for an experimental facility in Foxboro on Crack Rock Pond for ecological restoration of the upper Neponset River. The proposed purposes of this strategically located ecological facility included investigation of remedial techniques for Neponset Reservoir and the other reservoirs in Walpole, Norwood, Canton, and Sharon. Necessary facilities for improving fisheries would also be included. This ecological facility should be put high on the priority list.

Mouth of Neponset River in Dorchester Bay — The primeval inhabitants of Dorchester Bay and the land around the mouth of the Neponset River depended on fish and shellfish for sustenance. In 1630, the area now called Squantum was an important summer campground of the Massachusia people.

The first settlement of foreign colonists was in Mattapan. They subsisted largely on mussels, clams, fish, and nuts. For the first 5 years, fishing was very important to these settlers as the coastal land was unsuitable for farming. Cod was the basis of the first commercial fishery established in Dorchester before 1650.

The Neponset River estuary was fished by the colonists as early as 1634 when Israel Stoughton built a weir, and Swift and Clapp constructed a trap to catch migratory fish in 1681. The fishing was good. Records from the late 1700s recorded daily catches by one man on the lower Neponset River of 1500 shad in one day in June, and 2000 shad a few days later. The following year, he caught 80 barrels of shad in July, which he brought to Boston and sold. He also caught bass, menhaden, and even cod at this point. Striped bass and Atlantic tomcod were caught by the canoe-full.

In 1746, towns along the Neponset River petitioned the General Court to have fish ladders constructed at the several small dams already built across the river. Unfortunately, these did not work. In 1799, a new dam was constructed without a fish ladder, permanently blocking the migration of fish. There were riots in Canton by fishermen and a frenzy of activity by upstream communities to prosecute the dam builder, but they were unable to sustain the legal effort.

During the 1800s, alewives were so abundant in the estuary that they were smoked for family use and also shipped to market. The spring herring run up the East Branch through Canton and to Lake Massapoag in Sharon was particularly famous. However, by 1909 the river was completely dammed and badly polluted with industrial wastes, marking the end of the alewife and shad fisheries. Urban growth and filling of coastal wetlands also took a toll, and by the end of the 19th Century, commercial fishing in Dorchester Bay declined until only occasional sport fishing was practiced, as it is at present.

In 1941, the Department of Public Health closed some of the clam flats in Boston Harbor due to gross industrial and domestic pollution. After 1967, clams from the rest of the harbor had to be purified before marketing. Records from the state purification plant indicated that 7000 bushels of softshell clams from restricted Dorchester Bay flats were processed between 1959 and 1967.

The intertidal flats of Dorchester Bay contained 1000 acres of productive habitat for softshell clams in 1970. The large majority were grossly contaminated, and the remaining 13% were moderately contaminated. None of the areas were open to unrestricted clam digging. Seasonal digging by licensed diggers was permitted in restricted areas if the clams were then purified in the state facilities.

Thus, the primeval abundance of fish and shellfish in the Neponset River and Dorchester Bay was reduced to a contaminated and irrelevant remnant by the beginning of the 20th Century. The combination of small dams across the river and excessive industrial pollution eliminated the migratory fish except for a damaged smelt run in the estuary. Urban expansion and development decimated the near-shore coastal fisheries as well.

At the end of the 20th Century, fishing in Dorchester Bay and the Neponset River was restricted to recreational activity because of insufficient fish to harvest commercially. Clams were contaminated, and only 10% of the suitable habitat could be harvested, under strict supervision. The natural resources that had sustained the Massachusia people and also attracted the British colonists in the 17th Century, were virtually destroyed in three brief centuries, despite the millennia of evolutionary processes needed to develop them.

3.3.2 Charles River

The Charles River and its shores are major water and cultural resources for people in metropolitan Boston. For the last century, the Boston Marathon runners have been following the general direction of the Charles River from its beginning in Hopkinton to the Inner Harbor of Boston, finishing near the Science Museum, the Esplanade, and the berth of the U.S. Naval ship Old Ironsides in the Chelsea shipyards (Figure 3.17). The marathoners follow a more direct route 26 miles long, and some make the trip in less than 3 hours. The river meanders, however, thus its name of "Quineboquin" among the original Massachusetts people, meaning winding or twisting. In contrast to the marathon, the river travels 80 miles to get to Boston and usually requires about a month to make the journey.

Archaeological evidence from the Charles River shores indicates that a thriving society populated its banks since 5000 years ago. Fishing weirs and other artifacts showed that the natural resources of the primeval river were an important food source for these people.

More recently, the French explorer Champlain discovered the mouth of the river in 1604, confirmed by John Smith of the Plymouth Bay Colony in 1614. Smith was not an explorer, however, and it was not until Chief Squanto of the Massachusetts people offered to guide Governor Miles Standish up the river in 1622 that the white immigrants proceeded any further than the river mouth.

The first immigrant settler along the banks of the Charles River was the non-conforming William Blaxton who lived in isolation on Beacon Hill until 1630 when the Puritans drove him to an adjacent river valley that flows into Narragansett Bay. That river was subsequently named the Blackstone River after this religious refugee. His river is described in another portion of this book.

Figure 3.17 Map of Charles River Basin showing towns and Boston Harbor.

Engineers and developers immediately became active along the Charles River, building the first canal in New England from the Charles River to Mother Brook and the Neponset River, in order to power a grist mill in Dedham. About one third of the Charles River flow is still diverted by this canal. Eventually, over a dozen low dams were built across the Charles River to extract its energy (Figure 3.18).

By 1800, the Boston shore of the Charles River was partially filled with excavated material, creating Charles Street at the foot of Beacon Hill. Fifty years later, the orderly rectangular streets of Back Bay were laid out on additional excavated material. In 1890, a newly created sewer agency constructed combined sewers along the Cambridge and Boston shores of the river, setting the stage for a nuisance that continues a century later.

The combined sewer error was amplified by creation of the large scenic basin formed by construction of the Charles River Dam at the site of the Science Museum in 1910, as part of the plan for Frederick Law Olmsted's Emerald Necklace of parks around Boston. The dam was designed by John Freeman to eliminate the unhealthy and malodorous conditions documented from the first sanitary survey of the Charles River in 1894.

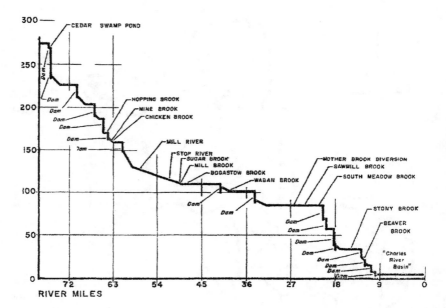

Figure 3.18   Profile of Charles River from Milton to Boston. There are also two salt barrier dams at the mouth of the river: the original Charles River Dam built in 1911 and the supplemental Warren Avenue Dam built in 1980. (From Smith, J.D., 1974. First Interim Report on Evaluation of Federal Program for Control of Pollution in the Charles River and Boston Harbor. Process Research Inc., Cambridge, MA. With permission.)

According to leading scientists at the time, the malodorous atmosphere around the tidal flats were the cause of malaria and other summer fevers. Fortunately, construction of the dam and flooding of the shoreline pools by the new reservoir did reduce the summer populations of mosquitos. These pests were responsible for transmitting sporadic epidemics of malaria and yellow fever brought from Havana, Cuba, by ships docking in Boston Harbor. These summer epidemics were the reason for the custom developed among New Englanders of taking their summers in the woods of Maine, rather than on their own beaches and rivers in Boston.

For a time, the new Charles River Basin and other portions of the river near the metropolitan area became a great source of enjoyment for the local population, reducing their tendency to flee in the summer. Several beaches were created upstream of the Charles River Dam, boat excursions and concerts became popular pastimes, and holiday events were established all along the banks of the lower basin.

Prior to World War II, several additional features were added, including Storrow Lagoon, intended to make the Charles River Basin the most beautiful and useful water park in America. The Boston Pops Orchestra began summer concerts along the Esplanade formed around Storrow Lagoon, including a Fourth of July celebration that has persisted for half a century, despite the foul condition of the nearby waters.

Clearly, creation of the basin to make the metropolitan area more healthy, more attractive, and more enjoyable as a natural resource was an honorable intention. However, this intention was satisfied for only a short time, if at all, because of the errors, poor maintenance, and mismanagement of the sewers hidden from the view of the area's eminent scientists, engineers, and philosophers.

The lack of correlation between the underground plumbing of the city and the above-ground intellectual keenness is an ironic and tragic failure of American engineers and environmental planners. Complaints and warnings from the plumbers and river crews who monitored these pipes and their smelly discharges never reached the ears of the Boston Brahmins. But a society that respects its philosophers more than its plumbers will end up with stinking rivers and harbors, as well as suspect philosophy.

The basin again became famous for its malodorous nature and as an open sewer, a century after John Freeman first tried to correct the problem by building the Charles River Dam. Shortly after World War II, a study by chemists from MIT determined that the basin was extremely polluted, containing both conventional and exotic contaminants of every variety. Some of these exotic substances came from combustion products of petroleum and included:

Alkylnaphthalenes
Alkylanthracenes
Dibutylphthalate
Di(2-ethylexyl)phthalate
Pyrenes

By Earth Day 1970, it was dangerous to fall into the Charles River in the lower basin; bridges and conduits were corroded by the acid and aggressive lower layer of the stratified waters. Sailors and crew members boating in the river had to be immunized against diarrheal diseases. The poor coxswains who won a crew race had to keep their eyes and mouths closed when they were ceremonially thrown into the river after winning a race. Imagine all this on the front door of the institutions that placed men on the Moon!

3.3.2.1 Upper Basin

The entire Charles River watershed encompasses 307 square miles. From an elevation above sea level of 500 feet along the southwestern rim, the drainage basin contains a large, central wetlands between Millis and Medfield covering about 31 square miles of the total watershed (Figure 3.17). These wetlands are a significant wildlife habitat and hydraulically dampen flood flows coming from the steeper tributaries and from upriver.

The upper portions of the Charles River received discharges from the Milford, Upper Charles, and Medfield sewage plants at the end of the 20th Century. Also, three state institutions discharged treated sewage to the Stop River, a major tributary joining the mainstem in Medfield. Below Medfield there are no further sewage discharges and the river is an attractive setting for canoeing and relaxation (Figure 3.19).

The middle and lower portions of the Charles River basin are within the domain of the Massachusetts Water Resources Authority, which conveys sewage to Deer Island and Nut Island in Boston Harbor for eventual treatment and discharge to the ocean. Thus, there are no significant sewage discharges to the Charles River downstream of Medfield. The main contaminants in this downstream and heavily

Figure 3.19   The South Natick Dam on the Charles River. At River Mile 41, between Dover and Natick, this is the last idyllic stop on the river before it enters the urban area proscribed by Route 128.

urbanized area are from stormwater and urban runoff, including debris washed off roadways.

The most heavily utilized portions of the Charles River are Storrow Lagoon and the Charles River Basin, between Cambridge and Boston, centers of cultural activity for the entire population of eastern Massachusetts. The Fenway and Muddy River parks designed by Olmsted border the basin on its southern shore.

The Charles River Basin is the site of innumerable regattas for sailboats and shells, including the Head of the Charles Regatta every autumn. On the Fourth of July, a quarter of a million people congregate around the Esplanade, on both sides of the basin, to hear the Boston Pops and their howitzer, church bells, and fireworks rendition of the 1812 Overture by Tchaikovsky. At the bicentennial concert of 1976, almost a million people lined the banks of the basin to watch the superb firework display. The basin is also lined by several major universities, including Harvard, MIT, and Boston University.

Because of its central role in the life of the people of Massachusetts and its discouraging problems with sewage contamination that repeatedly interfered with public use of several beaches on its shores, the state legislature sponsored sanitary studies in 1893, 1894, 1903, 1920, 1928, 1932, 1937 and 1939. This flow of water quality surveys was again resumed in 1967, followed by further studies in 1973, 1986, 1987, 1988, and 1990.

In terms of recent water quality, the last 20 years of concerted effort to improve the Charles River has only resulted in maintaining roughly the same unsatisfactory conditions that existed on Earth Day 1970. Human population growth and diffuse contamination sources have increased the amount of contamination in the basin as fast as new sewage treatment facilities have been constructed to treat the previous contamination.

In 1967, there were three major problems in the Charles River that severely limited its use for recreation. These problems were the low oxygen concentrations in the upper river due to excessive organic discharges, the large amount of algae in the middle and lower river that were repulsive and caused dangerous fluctuations in dissolved oxygen in the summer, and finally the enormous concentrations of exotic and offensive materials in the lower basin, from transportation systems, urban activities, and combined sewer overflows.

In the lower basin, these three problems were exaggerated by a fourth: a salt layer at the bottom of the basin that trapped the worst of the contaminants and slowly fed them back up to the upper, freshwater layer. This stratification increased during the 1970s due to increased leakage of the Science Museum Dam constructed in 1910.

Low Oxygen — The mean concentrations of dissolved oxygen in the upper river varied between 5 and 7 parts per million (ppm) during the summer of 1967, before the first Earth Day. The oxygen distribution was clearly related to the organic discharges, whose organic content is measured as the 5-day biochemical oxygen demand, or BOD5 (Figure 3.20). Most of the organic material came from partially treated sewage from the towns of Milford, Franklin, Medway, and Medfield. The BOD5 exceeded 5 ppm in the river below Milford, and reached 3 ppm below Medfield. Clean water has a BOD5 less than 1 ppm.

The largest organic loading in 1967 came from the municipal sewage plant of the Town of Milford, causing a noticeable drop in dissolved oxygen immediately downstream. In 1986, before the Milford sewage plant was improved, dissolved oxygen dropped as low as 2 ppm at river mile 72, due to the excessive organic material coming from Milford (Figure 3.21). Even after the treatment plant was improved, the dissolved oxygen remained significantly below the value of 5 ppm desired to protect fish, due to other discharges upstream.

Excessive Algae — A comparison of the dissolved oxygen throughout the Charles River over the 20 years from 1967 to 1987 showed the persistence of low oxygen conditions below Milford, and also showed that super-saturation of oxygen occurred in the rest of the river. Concentrations in the river as it passed through Sherborn, Dover, Natick, Newton, and Watertown reached as high as 12 ppm due to excessive photosynthesis by massive algae colonies. Unfortunately, this unstable condition also resulted in nearly complete depletion of oxygen on cloudy, hot days in late summer, due to the sudden curb on photosynthesis and the continuing respiration by algae and bacteria.

Unfortunately, the offensive algae population became more dense as the river entered the most heavily populated and potentially valuable portions of the river downstream of Newton. Measurements of the chlorophyll-a component of algae in 1973 reached peaks of 95 parts per billion (ppb) in Newton (Figure 3.22). Similarly,

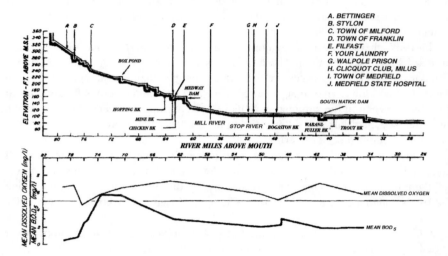

PROFILE of UPPER CHARLES RIVER

Figure 3.20   Dissolved oxygen and organic loading in the Charles River, 1967. The darker line in the bottom panel is the mean BOD5, the measure of organic concentrations. The letters A–J indicate the locations of the 10 wastewater discharges flowing in 1967.

in 1987, values of suspended solids, almost entirely algae, peaked at 40 ppm in Newton. These massive algae populations were due to the nutrients added to the river by the towns in the upper reaches. The total loading of these nutrients has not decreased sufficiently over the last 20 years to cause any appreciable improvement in the algae problem.

Diffuse Contamination — Material washed off of roadways, urban residential areas, railroad yards, and contamination originating from overflows of combined sewers have damaged the ecology of the lower Charles River for the last century. Although considerable expense has been incurred to separate the combined sewers in Cambridge and parts of Boston, the numerous roadways and railroads paralleling the river have probably compensated for improvements in the combined sewer overflows. In 1995, it was discovered that the major railroad yards in Allston had been discharging waste oil directly to the river for years.

Swimming in the lower river downstream of Route 128 has been prohibited during recent decades due to the poor clarity of the water, the high fecal bacteria values, offensive blooms of algae, and frequent oil slicks. In an extensive summer survey of bacterial contamination in 1990, it was found that combined sewer overflows and storm sewers were the source of most of this diffuse contamination. It was also concluded that even if all the combined sewer overflows were eliminated, stormwater would still bring in unacceptably high amounts of fecal contamination. The flow from urban sources between the Moody Street Dam and the Watertown Dam was indicated as a source of particularly high bacterial contamination.

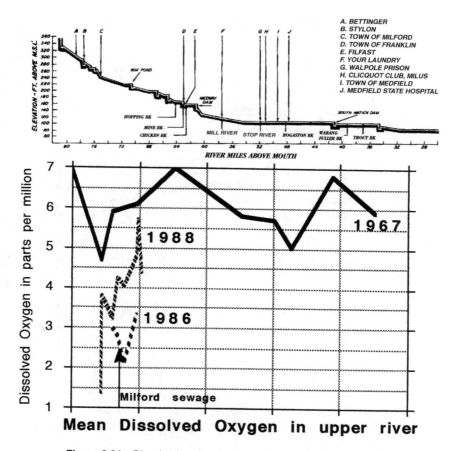

A. BETTINGER
B. STYLON
C. TOWN OF MILFORD
D. TOWN OF FRANKLIN
E. FILFAST
F. YOUR LAUNDRY
G. WALPOLE PRISON
H. CLICQUOT CLUB, MILUS
I. TOWN OF MEDFIELD
J. MEDFIELD STATE HOSPITAL

Figure 3.21   Dissolved oxygen in upper Charles River, 1967–1988.

Stratification above Science Museum Dam — Until about 1979, the bottom of the lower Charles River Basin, at the most scenic central point between Boston and Cambridge, was 60% seawater, due to leakage of the ocean through the old dam at the Science Museum, and to frequent operation of the large lock in this dam for small pleasure craft (Figure 3.23).

The stagnation of this lower salty layer often caused a depletion of oxygen, killing fish and releasing hydrogen sulfide to the upper layer, used by thousands of sailboats and rowing sculls. In 1978, the MDC (Metropolitan District Commission) installed air injectors in this lower layer to gradually inject oxygen, and in 1980 a new dam slightly downstream of the old one helped to gradually wash the saltwater back into the harbor. This $50 million investment caused immediate and noticeable improvements in the basin quality.

However, there was no provision in agency budgets for continued maintenance of the air injectors. They also curtailed monitoring of the oxygen in the basin once

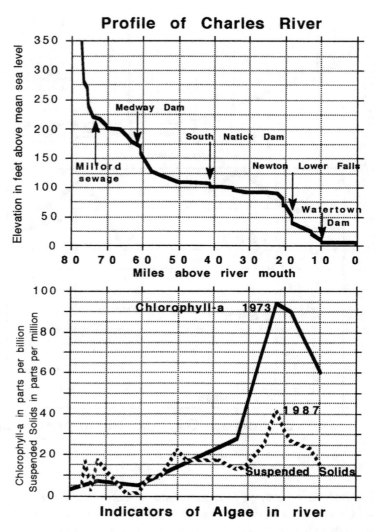

Figure 3.22 Impact of excessive nutrients in Charles River on downstream algae populations.

it improved. Also, they had operating troubles with the smaller locks on the new dam, and leaks began to develop in the new dam itself, allowing salt to intrude once again. Thus, when the air compressors wore out, no one noticed that the oxygen in the lower layer again decreased to zero as the salt again caused stratification.

However, the rowers on the basin noticed that the stainless steel cables that anchored their lane markers were corroding within weeks after installation, indicating highly corrosive conditions at the bottom. The salt, the hydrogen sulfide, and the acid conditions had returned.

This failure of the new system was largely due to political difficulty of the concerned agency in raising sewer and water rates. This had caused reductions in

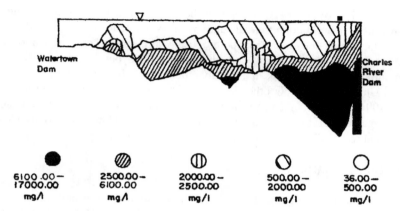

Figure 3.23   Saltwater stratification of lower Charles River Basin, 1970. Shaded areas indicate vertical distribution of salinity. Harbor water intruded through the Charles River Dam at high tide, reaching salinity similar to ocean water at the bottom, and causing stagnant and offensive conditions. (From Smith, J.D., 1974. First Interim Report on Evaluation of Federal Program for Control of Pollution in the Charles River and Boston Harbor. Process Research Inc., Cambridge, MA. With permission.)

their operating budget to pay for construction costs, but then did not allow for maintenance of the dam, the locks, or the air injectors. Approval of construction projects was easier to obtain than were operating funds. The lethargy of the responsible agency, the Metropolitan District Commission, was an important sign of the fundamental problem of improving the ecology of Massachusetts waters.

3.3.2.2 *Storrow Lagoon and Lower Basin*

Because of its magnificent setting and location among several large universities, the lower Charles River Basin and Storrow Lagoon have received more intellectual and research attention than any other part of the river.

Storrow Lagoon was designed for model boat sailing and general scenic enhancement of the basin, but the original intent of its designers has been frustrated by excessive algae, offensive odors, dark waters, and floating grease and oil.

Since Earth Day 1970, two proposals and pilot field studies have been conducted on new techniques for improving the water quality in Storrow Lagoon and in the Muddy River. The Muddy River is the second principal source of contamination of the Lower Basin, after the main flow of the river itself.

The proposals and pilot studies however have never been acted upon, a continuation of the lethargy and resistance from the water and sewer agencies responsible for the area. Despite benefits estimated to be between $7 million and $16 million per year for swimmable quality water made in 1975, the Metropolitan District Commission (MDC) did not act upon a proposal that was estimated to cost less than $4 million for construction and $400,000 annually for capital cost, operation, and maintenance. The benefit-to-cost ratio of roughly 10 to 1 was not enough to interest the lethargic MDC. They were the same group that decided a few years before to ignore the orders by the EPA to improve the treatment of their sewage going into Boston Harbor, even though federal funds were available to pay most of the cost.

In 1994, a pilot study and financial proposal was prepared for treatment of water quality in Muddy Brook, using conventional methods but placing the treatment system alongside of the brook in an arrangement that is being tried on the Calumet River in Chicago and a few other locations. The cost for this proposed treatment plant in 1994 prices is less than $3 million and it would go a long way in reducing the massive organic and fecal pollution coming into the parks and lower basin of the Charles River. Local responsible agencies have paid no attention to this proposal either.

3.3.3   Boston Harbor

Perhaps the most important aspect of the ecological deterioration of Boston Harbor is that attempts to restore it, beginning in the late 1600s, have shown a remarkably repetitive pattern of deterioration, poorly conceived attempts at restoration, and then deterioration again. The harbor and its drainage basin are in the third of these misbegotten cycles. Perhaps such repetition is human nature, and the reason King Solomon told us in the Book of Ecclesiastes:

What has been is what will be,
and what has ' een done is what will be done;
and th⌐   s nothing new under the sun.

Upon entering the haiᵤ   rea in the 17th Century and realizing the value of the site for their new settlement, the British colonists took advantage of the large natural supply of fish and shellfish for food. Estuaries and flats were harvested energetically to supply food for the growing immigrant community, eventually to be called Boston after their original port of departure on the east coast of England.

The major tributaries entering the harbor were also soon harnessed for power to turn mills, usually by constructing small dams. These dams were based on simple designs and experience gained from the British Isles, and were built by the immigrants with manual labor.

In this new immigrant colony, human wastes and wastewaters were thrown directly into the water, or deposited on land removed from the settlements. As the population grew, these practices produced obnoxious and dangerous conditions.

In 1630, the town of Boston was established on the Shawmut peninsula. Drinking water was obtained from a small well on the Town Common, soon replaced by a cistern and water conduit. As the population grew, the demand could not be satisfied from nearby sources that were becoming contaminated. After the War of Independence, water was drawn from Jamaica Pond, in about 1795.

In the 17th and 18th Centuries, it was assumed that diseases were caused by the bad odors emanating from this decaying organic material. Thus, in 1656, the citizens of Boston were ordered to throw their offal and garbage into the creeks so it would be washed away by the tides.

Gradually, the natural drainage system of creeks and swamps was supplemented with crude drains and sewers made of wood, which by the beginning of the 18th

Century led the offensive surface waters directly to the shores of the harbor or to the estuaries of tributary rivers. Flows from rainfall and tidal exchange diluted the most obscene collections of excrement and rotting carcasses, and also moved it further away from the now respectable homes of Beacon Hill and Dorchester, satisfying the rising esthetic requirements of the New Englanders.

3.3.3.1 Water Carriage of Wastes in Combined Sewers: Boston's First Ecological Mistake

The War of Independence and the establishment of a federal government occupied the elite of Boston for most of the 18th Century so no real advances in sanitary design were proclaimed until 1820 when indoor plumbing started to replace latrines and chamber pots for disposal of human wastes. However, the flush toilets needed a place to discharge large quantities of flush water along with the feces and urine. Thus, citizens were given legal permission to connect their toilets to existing storm drains in 1820.

The innovation of flush toilets also increased the quantity of water needed to supply homes; thus, larger amounts of water had to be withdrawn from the rivers, and reservoirs had to be built around the city to store water for use during the late summer dry season. This was the repulsive beginning of the concept of a water carriage system for sewage, in which human wastes are dumped into the clean flow from rainfall. For various reasons, this water carriage concept persists among engineers and planners to this day, despite its clear defects.

At this point, the sanitary planners were adding an additional dimension to the water-wasting habits of this new society that had now spread across the continent. The rising demand for large public water supply systems to supply not only drinking water and water for laundry and household cleaning — but also water to flush away excrement — became typical of American and European societies. It has not taken hold in Asia or Africa, except under Western influence. Over a century later, this fundamental difference in water use continues to differentiate Asian societies from those of Europe and the Americas.

Unfortunately, the American planners of 1820 were ignorant of what is now regarded by ecologists as a fundamental rule against mixing of waste and resource streams. In this case, the engineers were mixing the stormwater not only with the domestic sewage, but also with the industrial wastewaters — all in one combined sewer system. That basic mistake has haunted Boston and other American cities to the present day.

The water carriage system required increased supplies of water. Despite the fault in concept, the water supply engineers who developed these clean sources of pristine waters for the increasing populations began to reach a prominent position in American society. Masonry aqueducts and reservoirs full of pure water became objects of civic and engineering pride.

However, the sewers that carried away this water after it was contaminated were only celebrated while they were being constructed. Once they became tainted by their smelly cargo, they were difficult to discuss in polite society, and thus were soon forgotten. Fortunately, they were buried, so it was easy to forget them, and when they deteriorated or ruptured, it was a long time before anyone noticed.

3.3.3.2 The Small Dam Era

The Civil War between the industrial New Englanders and the cotton-producing South was preceded by rapid economic growth in the North that had an important impact on the waters of New England. In previous centuries, small wooden dams had created reservoirs on every brook and stream to power grist mills. Just prior to the Civil War, these dams gave way to masonry structures in order to provide large amounts of more reliable power for the cotton processing and spinning industries that grew out of Eli Whitney's cotton gin. In 1840, the Mills Act of the Massachusetts legislature gave the water rights away to the local industries, allowing them to dam as they pleased.

Along with the expansion of railroad systems along each major river, the multitude of small masonry dams provided an industrial network that carried New England to the end of the 19th Century on a wave of prosperity and growth. Enormous mill buildings began to appear on every river bank, especially near the estuaries where river flows were at their maximum and provided continuous power.

The cities also grew rapidly as the labor force increased to operate these mills and industries. And with the increased number of people came more sewage and consequent cholera epidemics. Hoping to prevent more cholera epidemics and to supply the expanding metropolitan populations with more clean water for domestic use, Lake Cochituate was tapped in 1848 to bolster the supply from Jamaica Pond.

3.3.3.3 The Dismal Sanitary Cycles of Boston Harbor

A review of recent history in Boston shows that sewage and water problems were always present, and that the solutions never stood the "test of time." These temporary solutions usually involved a large construction program. But the systems eventually failed to function, producing a crisis, and then another large construction program. Although many small crises had occurred in the early history of the city, the first major crisis was right after the American Civil War.

The First Crisis in 1865 — After a particularly lethal epidemic of diarrheal diseases in Boston in 1865, it was decided that the 100 miles of wooden sewers were not adequate to carry sewage far enough from drinking water sources. The sewers not only spilled sewage into rivers along their routes, but they were too small and discharged into the harbor estuaries, where tidal action could drive the wastes many miles upstream.

A fateful decision was made to entrust the solution for the rotting wooden sewers to the hydraulic engineers who had already been successfully employed in designing and building the systems for collecting pure water upstream. It was not surprising that these engineers proposed a solution that required construction of larger sewers and hydraulic basins. That was something they knew how to do. Unfortunately, this set a pattern of response to sanitary or ecological crises in the harbor that has been followed for over a century. And it has been followed not only in Boston, but throughout the Americas.

The First Public Works Project of 1884 — In response to the cholera epidemics, the failing wooden sewers were replaced with a larger system of sewers made of more durable masonry. By 1884, this Boston Main Drainage System was extended out to Moon Head at the north edge of Quincy Bay, where the sewage was held in masonry chambers during the rising tide and discharged on the falling tide. This simple hydraulic attempt at a solution to the problem continued as the sewage disposal system of Boston for a century, with little improvement. It was a hydraulic solution to an ecological problem, characteristic of the engineering approach adopted by Bostonians for their environmental and health problems.

The discharge of Boston's sewage to Quincy Bay was also an important historical precedent that Boston would regret a century later when the citizens of Quincy finally revolted. It was a court action by the City of Quincy that forced Boston into the present-day solution for its sewage problems. This solution, mandated by the courts, has turned out not only to be more expensive than expected, but it also has some basic ecological flaws.

At the end of the 19th Century, the "great sanitary awakening" was stimulated by the unhealthy and offensive conditions of the waters immediately around the city of Boston. The scientific basis of this "awakening" was the germ theory of disease, and the discovery that mosquitos transmit malaria and yellow fever. The previous theory — that they were caused by bad airs — was discarded, and sanitarians began in earnest to plan more healthy cities, based on their new knowledge. The stimuli for all this heady intellectual and engineering activity, however, were the filthy muck and the pestilences of the industrial centers spreading across the drainage basin and shores of Boston Harbor like a fungus.

At the same time that they were building the first Main Drainage System for Boston, the hydraulic and dam engineers were making great strides in increasing the upstream supply of domestic water. The Massachusetts Board of Health was created in 1864 to investigate the decreasing quality of the suburban rivers that supplied Boston with drinking water. In 1872, the Sudbury River was dammed to provide additional water, and the Metropolitan Water District was established in 1895 to find even more pure water. It began construction of Wachusett Dam on the pristine upper reaches of the Nashua River, along with an aqueduct to bring the water to Boston (Figure 3.24).

As part of the continuing program of civil works, in 1889 the Metropolitan Sewerage District was created to expand the sewers of the Main Drainage System. Construction of these sewers, following the tributary geography, resulted in a North Drainage System along the Charles and Mystic Rivers, and a South Drainage System along the Neponset River.

The first survey on water quality and contamination of bottom sediments in Boston Harbor was conducted in 1898 after construction of the North Main Drainage System. Unfortunately, this new sewer was already causing visible degradation of the harbor bottom. A dark, 700-acre area of contaminated sediments was discovered on the harbor floor, at the point of the main discharge. A similar survey 70 years later showed that the sludge deposits were worse in quantitative terms, and that they were concentrated in the Inner Harbor.

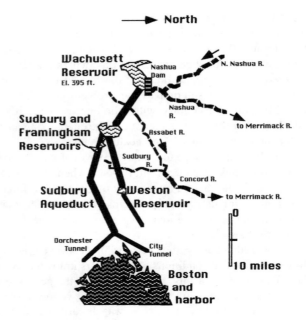

Figure 3.24 First stage in Boston's water supply system, 1900.

Nonetheless, the sewers were a major public works effort, and were thought to be a modern and final solution to a problem that had persisted since colonial times.

The Second Sanitary Crisis in 1919 — There were some unfortunate problems with the masonry sewers and the Moon Island Head Works created by the new Metropolitan Sewerage District. These sewers and outfalls were found to be a poor and temporary solution as the waste was soon returned to the shores of Boston by incoming tides. The sewers were also hard to maintain, and frequently overflowed because they carried storm water as well as sanitary wastes. Before World War I, conditions around Boston once again had become so offensive that another great campaign was conceived as the final solution.

After World War I, Boston and the surrounding communities organized the Metropolitan District Commission (MDC) to deal with the environmental problems that had accumulated, to provide clean water for the expanding population, and to improve the parks and green spaces around the metropolitan area.

In 1919, sewage pollution had caused the state Board of Health to close several clam beds in the harbor. The MDC assumed control of all the sewers to try to fix the problem. However, the root causes of the problems — the combined sewer overflows — were not dealt with, and by 1939 all clam beds in the harbor were contaminated. In 1940, it was recommended that all three sewage discharges into the harbor receive treatment.

That same year, Winsor Dam was completed on the Swift River in central Massachusetts, creating Quabbin Reservoir on rivers that had previously been part of the Connecticut River Basin. This enormous reservoir was supplemented with

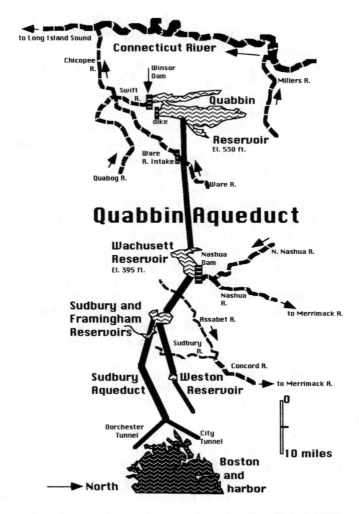

Figure 3.25 Water supply system for metropolitan Boston, 1996.

draws on the nearby Ware River (Figure 3.25). The waters derived from these lightly populated watersheds were then delivered to Wachuset Reservoir to supplement its storage for supplying the MDC system. Quabbin Reservoir was so large that it did not fill to capacity until 1946. This source has supplied Boston since that time, even with the loss of the Sudbury River Reservoir due to contamination of the upstream tributaries of the Sudbury River.

The completion of Quabbin Reservoir and its complementary system of aqueducts, tunnels, and intermediate reservoirs was a source of great pride for dam builders and water supply engineers in Massachusetts, one of the largest hydraulic systems constructed solely for domestic water supply. It brings clean water from pristine rivers and lakes containing abundant fish, snails, and mussels, supplied by forested land inhabited by deer, eagles, and bears, through 65 miles of large concrete and steel aqueducts, across three major river basins, eventually reaching homes and industries along the coast of Massachusetts Bay. By the year 2000, almost all of that water will then be discharged through the system currently under expansion, into Massachusetts Bay near Stellwagen Bank, home of the nearly extinct right whales.

This grand traverse of the water across Massachusetts from the home of small snails to the bay of whales has a certain lunacy to it that requires some imagination to understand. What happens to this water after it is so majestically transported these 65 miles, at such expense? How much is used to quench the thirsty appetite of the 2.5 million New Englanders? With such a grandiose effort, bringing the water so far across these intervening river basins and communities, one would imagine all of this precious water is for human consumption. What else would justify the expense?

But no! Not all of the water extracted from the Swift and Ware River Basins is used for drinking water. If one calculates that 2.5 million people drink 8 glasses per day, that would be a direct human consumption rate of less than 7 million gallons per day. Of course, a significant amount of pure water is also needed for bathing, kitchen use, clothes washing, and other high-quality household uses. For the average person, all of these needs require about 32 gallons per day, or 80 million gallons for the entire metropolitan population. However, the magnificent reservoir and aqueduct system is designed to safely deliver 300 million gallons per day. Thus, the amount that needs the purity of the Quabbin Basin is only 26% of the total volume supplied to the metropolitan area! Then why is so much water taken from Quabbin?

Precious Water Wasted to Flush Toilets — The problem started in 1820 when the Boston Brahmins decided to allow flush toilets in their fair city, and connected them to the primitive storm drains that had been previously constructed by innocent drainage engineers.

Believe it or not, a large portion of the pure and expensive waters of Quabbin Reservoir, transported across 65 miles of Massachusetts hills and valleys, is used to flush human feces and urine into Massachusetts Bay!

Imagine using expensively collected and laboriously transported super-pure water to flush feces down a pipe — the marvelous design concept known as the water carriage sewerage system!

This is not the only strange activity in the metropolitan water system of Boston. A large part of this Quabbin Reservoir water is also used to water the grass around suburban homes in the hot, dry summers. And, of course, cars must be washed and gardens watered. And then industries need water to flush away metal-plating wastes and wash spent fibers from textile and paper mills into the sewerage system, also leading to the whales in Massachusetts Bay.

There is a certain hydraulic logic in this system of water carriage; but there is no common sense in it, despite the legacy of Thomas Paine and other New England revolutionaries. It is a form of ecological insanity that is kept hidden, buried underground in steel aqueducts and concrete sewers, finally to surface near Boston Light, or someday out toward Stellwagen Banks where only the whales, cod, dolphins, and tuna will see it.

Thinking carefully, one would see two separate ecological faults in the water and sewage system of Boston and the rest of the Americas. First, pure water is too precious to be used for purposes such as transporting feces, watering grass, and flushing away spent fibers. Second, disposal of the toxic elements such as heavy metals, industrial solvents, and other dangerous materials should not be combined with the relatively innocuous organic wastes from the human body, if the wastes are to be discharged into rivers or oceans. Let us ponder on these errors and search for a rational system for our water and sewage.

The Second Public Works Project in 1952 — World War II diverted Boston's attention away from sanitary problems, but eventually the MDC constructed sewers and sedimentation basins on the harbor discharges. The tanks were constructed at Nut Island in 1952 and Deer Island in 1968, thus slightly reducing the organic and fecal material in the discharges. The Moon Island discharge was closed and the discharge connected directly to Deer Island, except for spills to the harbor in frequent emergency situations.

In the simple treatment facilities at the two points of discharge, the sewage was captured in the sedimentation basins during the incoming tide, and then the stored flow and settled sludge were released to the harbor on the outgoing tide. Construction of the tunnels, pumping stations, regulators, and sedimentation works utilized the combined talents of the city's finest civil engineers. The finished works were proudly acclaimed as evidence of Boston's place as the most modern of cities, with the finest sanitary works in America.

The proud moments were short, however, lasting only as long as the pipes and tanks were new and unsullied. Once they began to carry sewage, public interest turned to other, brighter things.

The Nut Island and Deer Island facilities seemed quite a sensible solution to the MDC, except for the complex hydrodynamics of the harbor. Unfortunately, the tidal flux in most of the Inner Harbor and Dorchester Bay is quite small, and the material was actually not transported to the ocean as hoped. Instead, much of it was dispersed, mixed, and frequently returned to the point of discharge. In fact, there were currents that took some of the material from Deer Island and Nut Island upstream into Dorchester Bay, depositing solid wastes along the shore.

This deposited waste piled up on top of the material still being discharged directly along the shoreline by malfunctioning or unconnected sewers, which continued to proliferate. Although a great deal of money and effort had been concentrated on building the major tunnels and treatment works, little money was left for their operation and maintenance. Also, the city sewers were now ancient and crumbling, and many of them leaked directly into the harbor. At high tide, the ocean penetrated the sewers through the same leaks, resulting in transport of the sewage up the pipes instead of downstream to the island outfalls.

> In the colonial era, the residents of Boston threw their slop into the Inner Harbor. Three centuries later, with technological advances, they threw their slop a little further — into Dorchester Bay and the Outer Harbor. Given the intellectual resources of Boston, this was hardly an elegant solution.

A second major source of contamination of the harbor was the combined sewer system. The pipes were designed to let large storms overflow through regulators into the natural water bodies, including Storrow Lagoon, the Inner Harbor, Chelsea Creek, Dorchester Bay, and the many lagoons and ponds in the Emerald Necklace around Boston designed by Frederick Law Olmsted. Although this was a satisfactory way to limit the rate of flow into the sedimentation facilities, it had the negative impact of frequently discharging an offensive mix of fecal material, industrial wastes, and waste oil into the recreational areas of the city, and into the Inner Harbor. This combined sewer system has been a major headache for centuries. By the time the Deer Island sedimentation basins were in operation, their impact was drowned by the raw sewage spurting out of the plumbing leading to Deer Island and Nut Island.

The sediments in the harbor in 1967 were thick black sludges that contained large amounts of heavy metals, petrochemicals, organic carbon, and organic nitrogen. Most of the harbor bottom was covered with sludge deposits by this time. The deposits accumulated because the wastes entering the harbor exceeded the waste dispersion capacity of the tidal currents and river flows. In many locations, the sediments emitted hydrogen sulfide gas, indicating excessive organic matter and oxygen depletion.

Sludge deposits in the Inner Harbor were over 3 feet thick in 1967, but accumulations tended to be less toward the ocean and in areas where tidal currents were large. This pattern was reflected in the distribution of polychaete sludge worms. The worms live in over-enriched sludge deposits of decaying organic matter. Although the worm populations were somewhat inhibited in the Inner Harbor by heavy metals and toxic materials, the density of polychaete worms exceeded 5000 worms per square foot at the outlet of the Inner Harbor (Figure 3.26). (Worm population densities greater than 200 per square foot indicate overly enriched sediments.) The Inner Harbor, Dorchester Bay, and the inlet west of Deer Island had worm densities between 1000 and 5000 worms per square foot. Only Nantasket Roads, the secondary shipping channel where tidal currents are strong, had low sludge worm populations.

Accumulation of heavy metals and other toxic materials had also reached excessive concentrations in the harbor sediments. The metal concentrations were highest in the Inner Harbor, around Deer Island, and along Dorchester Bay. Lower concentrations were found in the less industrialized areas in the southern portion of the harbor. The concentrations of zinc, copper, lead, cobalt, cadmium, chromium, vanadium, molybdenum, and mercury all exceeded the safe levels established by the EPA to protect marine organisms, usually by a very large factor (Table 3.3).

Large numbers of bacteria and other organisms causing human diseases were found in the harbor in 1967. The concentrations of coliform bacteria, which indicate human sewage, were high throughout the harbor. Softshelled clams were contaminated by these organisms, about 6% containing agents of typhoid fever. Clams also contained pesticides at concentrations toxic to marine organisms.

Faulty operation of the new MDC system began within a few years after it had been constructed because the system requirements soon surpassed the budgets dedicated to operation and maintenance, although initially there seemed to be little need

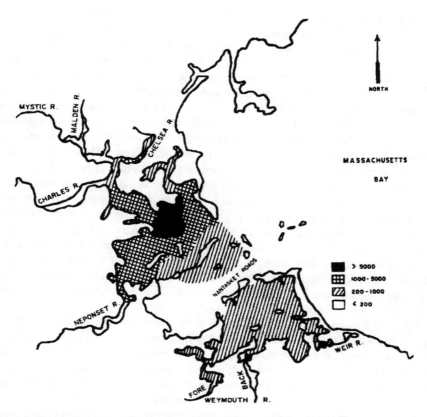

Figure 3.26 Accumulation of sludge worm populations in sediments of Boston Harbor, 1967. (From Smith, J.D., 1975. The Impact of the Federal Water Pollution Control Act on the Charles River and Boston Harbor. Process Research Inc., Cambridge, MA. With permission.)

Table 3.3   Heavy Metal Concentrations in Sediments of Boston Harbor, 1972

Toxic heavy metal	Maximum concentration observed, in ppb	Maximum acceptable concentration, in ppb according to the EPA
Zinc	1360	100
Copper	494	50
Lead	675	50
Cobalt	37	
Cadmium	29	10
Nickel	87	100
Chromium	433	100
Vanadium	1110	
Molybdenum	154	
Mercury	6	1

for expenditures on the new system. Gradually, it deteriorated but no one was interested in spending the money required to keep it operating as designed, or to improve it as the flows and loads expanded with the growing population.

After the First Earth Day in 1970, state and federal governments began to urge Boston and the MDC to separate the combined sewers, repair the ancient sewerage systems around the harbor, and improve the degree of treatment of the effluent discharged to the harbor. Assistance with the cost of planning and construction of these improvements was offered by both federal and state authorities to the cities, towns, and the MDC as inducements. The local government agencies were expected to pay only 10 to 20% of the planning and construction costs. However, they were not offered any financial help with the operation and maintenance costs.

The lack of money for operation and maintenance is a defining characteristic of sewerage systems in Boston, and in most cities. Perhaps the buried sewers are never visually evident, and thus they are forgotten. Perhaps the people who are placed in charge of operation and maintenance have a difficult time entering the chambers of finance because of the smell on their clothes. Whatever the reason, it is one factor that has remained constant throughout the cyclical history of the rise and decay of sanitary facilities in Boston. The phenomenon is quite universal. Despite this, it has always been ignored by planners and public agencies when responding to a sanitary or environmental crisis.

Most of the smaller cities and towns in New England had agreed to several aspects of the Clean Water program and slowly began separation of the sewerage systems, treatment of the storm overflows, and improved treatment of their sewage effluents. However, Boston and the MDC agreed only to an agonizingly slow program to separate the sewerage systems. Also, the cost of improving the Deer Island and Nut Island facilities seemed too high for the Bostonians.

By this time, the industries in the metropolis had expanded in number and variety, and were also discharging their wastewaters into the combined sewer system. Thus, the wastes reaching Deer Island sometimes contained toxic materials that retarded or killed the bacteria involved in the digestion process. Often, the month of detention in the digestors did little to improve the sludge. The industrial wastes were grossly offensive and also toxic to aquatic or marine life.

Industrial pollution in Boston Harbor became the most severe in the nation, indicated by 1984 concentrations of a very toxic class of contaminants, the polychlorinated biphenyls (PCBs), which reached 17 parts per million (ppm) in the Boston Harbor sediments, compared to 0.1 parts per billion (ppb) in the clean waters of Casco Bay in Maine, or 0.05 ppb in Chesapeake Bay. Fecal contamination was similarly excessive, giving Boston Harbor its unenviable first national ranking in the 1988 presidential campaign as the dirtiest American harbor.

As fast as the new sewers and overflow treatment facilities were constructed, older sewers and overflows decayed because of the lack of maintenance. Tide gates rusted open, allowing sewage to flow into the harbor and saltwater to flow back into the sewers on a regular tidal cycle. Within a few years after completion of the Deer Island treatment works, the contamination of the harbor had again returned to its most dismal condition. Clam flats and beaches were closed because of contamination. Summer storms caused large burps of sewage onto the beaches.

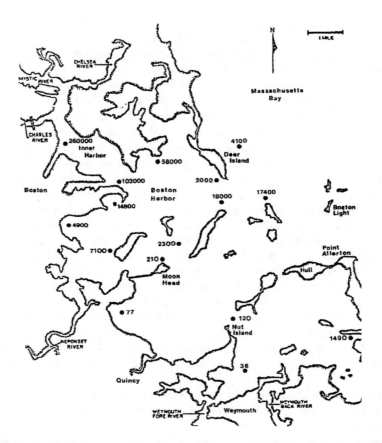

Figure 3.27 Numbers of fecal coliform bacteria found at various locations, in 100-milliliter samples of water from Boston Harbor during the summer of 1972. Although this survey was made soon after the Deer Island sedimentation tanks were completed, the fecal contamination had again become excessive, especially in the Inner Harbor. (From Smith, J.D., 1975. The Impact of the Federal Water Pollution Control Act on the Charles River and Boston Harbor. Process Research Inc., Cambridge, MA. With permission.)

There were 21 public beaches operated by the MDC in Boston Harbor by 1980. Before the MWRA began its current cleanup in 1985, 13 of those beaches were closed in a typical summer, due to gross fecal contamination. Most of it came from the malfunctioning sewerage system, and some from the Nut Island discharge.

Fecal contamination is measured by the number of coliform bacteria. They exceeded a quarter of a million per 100 milliliters of water in the Inner Harbor in 1972, almost the equivalent of raw sewage (Figure 3.27). Concentrations of coliform bacteria off the Dorchester beaches ranged from 5000 to 15,000 per 100 milliliters, much higher than the desired values, which should be less than 200 per 100 milliliters.

Only half the clam beds around the harbor could be used, even in dry weather. After heavy rains, none of them could be harvested, resulting in an annual loss to clam diggers of about $4 million under the MDC operating system.

Because of the overwhelming evidence of harbor degradation, it was expected that the completion of the sedimentation facilities at Deer Island in 1968 and the tidal storage tanks at Nut Island would cause marked improvements in all aspects of Boston Harbor water quality. But they did not.

Important indications of the reason for the seemingly ineffectual impact of these early Nut Island and Deer Island facilities can be seen in the pattern of sludge deposition, which is highest in the Inner Harbor, and in the water clarity, which was lowest at mouths of the major rivers. Together, these two indicators showed that the primary sources of degradation in the harbor were near the river mouths and along the city shoreline, not at the discharge points from the sedimentation tanks.

The harbor waters had high clarity and low turbidity near the Deer Island discharge points. Clarity of water decreases as measured turbidity units increase. Measurements of turbidity throughout 1967 showed that the waters near Deer Island were almost as clear as the open ocean value of 3 turbidity units, whereas the turbidity at the mouths of the major rivers went as high as 23 turbidity units (Figure 3.28). This was a strong quantitative indication that the major sources of harbor degradation were not the Deer Island discharges, but the discharges in the inner portions of the harbor.

The main sources of harbor contamination were the colonial legacy of combined sewer overflows and the metropolitan disregard for maintenance of the existing sewerage system. In the inner portions of the harbor along the city shoreline, there were over 200 raw sewage discharges in 1967, due to broken and ancient pipes (Figure 3.29). No matter what degree of treatment was provided at Deer Island, these myriad direct discharges to the Inner Harbor would continue to damage the harbor ecology.

In 1967, there were over 100 combined sewer overflow outlets, most of them with malfunctioning regulators. These sewers were broken, full of sediment, and in a generally chaotic state. During high tide, the ocean water often flowed into the system as far as the treatment facilities at Deer Island. This addition of saltwater prevented the sludge digestors at Deer Island from functioning, and hydraulically overloaded the facility. There were also dozens of stormwater sewers that often carried human sewage due to illegal connections. The 10 permitted discharges included the MDC and MWRA facilities, as well as a few industrial discharges.

A confirmation of the importance of the Inner Harbor discharges was seen in the distribution of coliform bacteria in the harbor in 1972, long after the Deer Island and Nut Island facilities were in operation (Figure 3.27). The Deer Island and Nut Island facilities clearly had not improved the contamination of public beaches, one of the major concerns of harbor residents.

Thus, by Earth Day 1970, it was clear that the problem with Boston Harbor was the malfunctioning and poorly maintained sewer system. Unfortunately, there were no funds available from federal or state Clean Water programs to assist the MDC or the city of Boston in correcting these problems. The MDC sponsored a study on the Dorchester sewers that recommended increased maintenance and more rational operation of the system; however, it was not implemented. There were construction grants available through the EPA for construction of a new treatment plant at Deer Island, but MDC and city officials resisted the temptation, knowing it would not

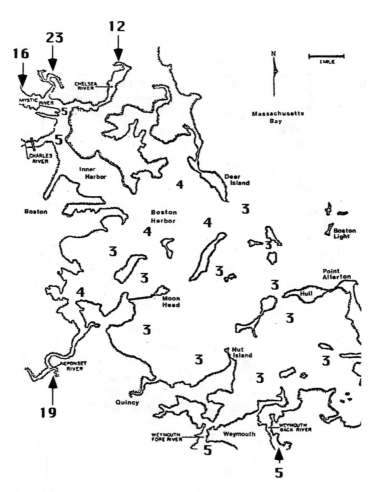

Figure 3.28   Turbidity of water in standard units in Boston Harbor, 1967. Arrows at river inlets indicate turbidity in river flow. (From Smith, J.D., 1975. The Impact of the Federal Water Pollution Control Act on the Charles River and Boston Harbor. Process Research Inc., Cambridge, MA. With permission.)

solve the problem. However, they never found money to deal with the real problems either, so conditions went from bad to unbearable.

The Third Sanitary Crisis of 1980 — By the 1980s, the few remaining fish in Boston Harbor were in serious trouble. In 1985, winter flounder in Boston Harbor had a 42% prevalence of liver cancer, the highest on the East Coast, and an alarming increase over previous years when the prevalence was 8%. The prevalence of fin rot among winter flounder in the harbor was 44%. Flounder live on the bottom of the harbor, and these diseases reflect the contamination of the sediments with heavy metals, PCBs, petroleum residues, and fecal material. Almost 5000 acres of shellfish habitat were closed to clam digging in Boston Harbor in 1985 because of bacterial

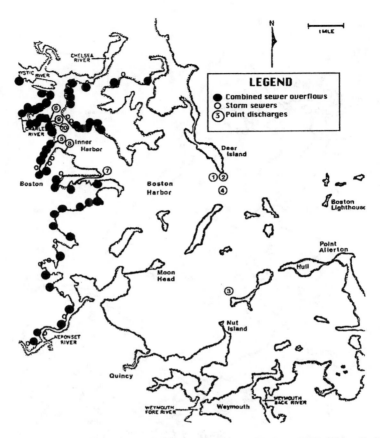

Figure 3.29   Locations of sewage discharges into Boston Harbor, 1967. (From Smith, J.D., 1975. The Impact of the Federal Water Pollution Control Act on the Charles River and Boston Harbor. Process Research Inc., Cambridge, MA. With permission.)

contamination. This marked the "third sanitary crisis" for Boston (Figure 3.30). Even our crude analysis showed that these crises were becoming more frequent. They should not be allowed to continue!

Once again, the citizenry complained to government officials, but to no avail. The citizens knew that there was a federal regulation that all ocean discharges of sewage must receive secondary treatment, but the two MDC discharges at Nut Island and Deer Island did not even meet the requirements for primary treatment.

Conditions had became so obnoxious in the early 1980s that the federal court reluctantly intervened, following a suit from the people of Quincy who could no longer tolerate the large clumps of fecal matter on their beaches. Using the simple enforcement tools available — but not a lot of foresight or historical review — the federal court ordered Boston and the MDC to meet the federal regulation about secondary treatment and to repair and expand the broken sewerage system.

This court action was understandable in light of the intolerable conditions, but the legal constraints resulted in a solution that contained serious flaws. Instead of a

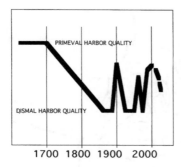

Figure 3.30 Symbolic portrayal of the cyclical history of ecological quality of Boston Harbor since the British immigrants arrived in the 17th Century. The quality of the harbor is approaching the downward turn into its fourth cycle, due to the increasing costs of operation and maintenance.

careful and thorough ecological analysis of the needs of Boston Harbor, the court had to fall back on the federal EPA regulation requiring secondary treatment. The flaws in this ruling are that the degree of treatment given to its sewage had little to do with Boston's water problems, and that the money spent on secondary treatment would thus be unavailable to correct the basic problem — which had always been the operation and maintenance of the pipes and structures in the sewerage system.

Dire Prediction — As the financial resources of the Boston metropolitan area are poured into construction and operation of a large secondary treatment plant, there will again be no money left for operation and maintenance of the sewers and tide gates leading to that plant, and conditions in the harbor will once again deteriorate into an ecological crisis.

To carry out this new task, the Massachusetts Water Resources Authority (MWRA) was created in 1984, and obediently began the planning and construction of a new treatment facility at Deer Island that would also take the flow from Nut Island. After giving the sewage secondary treatment, the new system of pipes would carry it about 10 miles offshore, deep into Massachusetts Bay (Figure 3.31). The MWRA also began to use its capital funds to renovate the sewerage system, including expansion of major sewers and construction of new tunnels to adequately handle present and future flows. The MWRA program was promised congressional support and finally received grudging support of the local and state agencies as well.

One of the theoretical reasons given by the MDC in resisting the original orders to provide secondary treatment was that it would be a misuse of funds because the proposed system was actually overdesigned. This was theoretically correct because the MDC and the city of Boston had caused the deteriorating conditions in the harbor by neglecting operation and maintenance, not because they had not provided adequate treatment at Deer Island. This inaction by the MDC and the city of Boston in the 1960s and 1970s had set the stage for the latest crisis and the latest public works project of the MWRA. This will be an expensive disaster that will gradually unfold about the time the $6 billion project is completed, and no funds are available to operate and maintain the sewerage system.

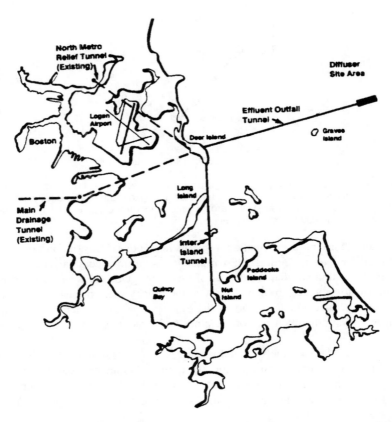

Figure 3.31   Proposed Massachusetts Water Resources Agency (MWRA) sewers and tunnels, to be completed about the year 2000. (From Smith, J.D., 1975. The Impact of the Federal Water Pollution Control Act on the Charles River and Boston Harbor. Process Research Inc., Cambridge, MA. With permission.)

The Third Public Works Project of the 1990s — Following the creation of the MWRA, fairly rapid progress was made on construction of the new, larger sewers and tunnels, and the Deer Island facility. Water quality in Boston Harbor slowly improved as overflows decreased and large pipe breaks were closed. By 1996, beaches began to be safe for summertime use; the health of fish in the harbor began to improve; clam beds were reopened; lobster, flounder, and even harbor seals returned; and visual evidence of contamination began to diminish. The clarity of the water in the Inner Harbor increased from 6 feet to a depth of 8 feet. These improvements all occurred long before the Deer Island facility and its deep ocean outfall were completed.

In the early days of the MWRA when funding was easy, a farsighted program was initiated to reduce the amount of toxic materials coming into the sewer system, especially heavy metals. The MWRA also intensively monitored the quality of the harbor ecology as each new component of the system was put into operation. In the early years, the harbor responded well. In this heady decade, the harbor porpoises returned and the numbers of unsightly tumors on flounder decreased.

These positive signs all appeared before the secondary treatment facility began to function at Deer Island, and while the sewage was still being discharged at the harbor mouth. By 1996, there was a feeling of elation by harbor users, even though the treatment plant and outfall were not to be finished for another 3 or 4 years.

No Need for Secondary Treatment — This early improvement in the harbor should be understood as an important guide for evaluating future activities of the MWRA. Secondary treatment was not necessary to obtain these improvements. All that was needed was plumbing in good repair.

In 1993, the prevalence of early liver diseases in winter flounder had decreased to 45%, falling far below the maximum of 75% in 1987. Liver tumors had almost disappeared by 1993. Herring runs in Weymouth Back River in the southwest corner of the harbor reached as many as 600,000 per year in 1991. Although the numbers fluctuated considerably, they were much higher than the numbers observed in 1970.

Contamination of harbor lobster and winter flounder with exotic industrial chemicals such as mercury and PCBs showed significant decreases between 1986 and 1990, and the concentrations of PCBs and combustion byproducts in Deer Island mussels decreased more than 50% during that same period. This improvement occurred before the Deer Island plant even had adequate primary treatment in operation. Would secondary treatment really be necessary to restore the harbor ecology?

In the proposed MWRA system, expected to be completed by the year 2000, it had been decided to discharge the treated effluent about 10 miles outside of the harbor entrance in a large undersea pipe, with considerable diffusion into the 150 feet of ocean at the point of discharge. In terms of the simple measures of organic pollution, such as dissolved oxygen and suspended solids, this amount of dilution in the deep ocean would reduce impacts to almost unnoticeable magnitudes, even without treatment of the discharge. Thus, as long as the sewer pipes are kept in good shape, the harbor would maintain its improved quality. This was already becoming a reality in 1996 while the MWRA had capital funds available for repair and renewal of the plumbing, and before the operating and treatment expenses for the Deer Island facility became significant.

Some problems could continue, even with the deep ocean outfall, because of toxic industrial wastes. Heavy metals, industrial solvents, petrochemicals, and other exotic contaminants caused severe damage to the harbor resources for a century, and would also damage the deep ocean resources, even with secondary treatment. In 1984, over 3000 pounds of toxic heavy metals flowed into the Deer Island and Nut Island treatment plants every day. However, the MWRA source reduction program for toxic wastes caused a marked decrease in the amount of heavy metals being discharged, and has probably also reduced many other toxic materials. By 1992, less than 700 pounds of metals were reaching the MWRA plants. To lower this amount further, industries must conserve, recycle, and otherwise reduce their wastes — the most sensible way to control these exotic and toxic substances.

Fundamental Faults and Excessive User Charges of the MWRA — Despite the general satisfaction over the recovering harbor, there was a fundamental fault in the program ordered by the federal court and implemented by the MWRA, which will soon return to haunt New England. In the usual excitement about the enormous construction project, no one had remembered the history of the Metropolitan Sewerage District, nor the eventual failure of the MDC.

No one seriously considered whether the MWRA would repeat these failures and similarly be unable to afford operation and maintenance of the new facilities.

Only planning and construction grants are available from the federal government. Operation and maintenance costs have to be borne by the MWRA by collecting water and sewer charges from its ratepayers, the people of Boston and the metropolis. Concrete fever, the same malady that had blinded sanitary and environmental planners during previous "sanitary awakenings" in New England, caused the planners' vision to blur and their brains to forget the need to operate and maintain these wondrous hydraulic creations.

The MWRA was already in deep trouble by 1996 due to financial constraints. The dismal history began to repeat itself by the centennial of the Boston Marathon, 3 years before the Deer Island plant was to start secondary treatment and the deep ocean discharge. Because Boston had waited too long to respond to the federal assistance offered in the Clean Water Act of the 1960s, construction aid from the U.S. Congress proved difficult to obtain.

With political changes in Washington, the financial assistance expected from the federal government began to shrink, forcing the MWRA to increase its charges to users even before their new facility began operation. Skyrocketing rate increases to cover costs of construction forced the authority to start cutting back on staff and programs, and there was little hope for future assistance.

Approach of the Fourth Crisis in Boston Harbor — A severe crisis will occur when the MWRA secondary treatment plant is in full operation, about the year 2000. At this point, the MWRA ratepayers will have to cover not only the capital costs, which are rising at 10% per year, but also the completely new operational costs for electricity, chemicals, and skilled technicians and engineers to run the new facility. Furthermore, the MWRA will have to start repairing the sewers and tide gates they installed several years ago. The MWRA will not have sufficient income to cover those costs.

This fourth crisis is part of the repeating tragedy of Boston Harbor. The expected crisis at the end of the century occurred three times before in Boston Harbor, in a very familiar pattern (Figure 3.30). Enthusiasm for large public works was quickly replaced by lack of support for sustained operation and maintenance. Temporary improvements in water quality removed the political pressure for continued action, and then the projects were slowly abandoned in the face of escalating costs.

Not many people noticed the previous lapses for a decade or two because the problems were mostly underground and related to human excrement in old sewer pipes, a subject somewhat delicate for Bostonians to deal with in the absence of a crisis. Thus, the crusty New Englanders are in danger of sliding into a fourth repetition of their cyclical drama of Boston Harbor.

The financial aspects of this approaching crisis are clearly evident in reports of the MWRA and their financial supporters. From former annual sewer and water

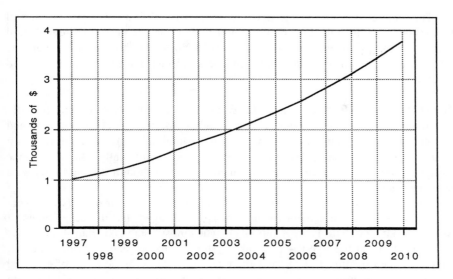

Figure 3.32   Massachusetts Water Resources Agency projected annual water and sewer costs per average household in thousands of dollars, 1997–2010.

rates per household of $100, charges tripled by 1985 when the MWRA took over the system. By the time the facilities start operation in the year 2000, it is expected that annual sewer charges per household will skyrocket to $1200 (Figure 3.32). For many households, this is equivalent to their annual property tax.

A statewide taxpayers' revolt in Massachusetts during the previous decade limited property tax increases to a growth of only 2.5% per year. If they objected to such a small increase in their property tax, will the MWRA ratepayers be willing to pay 10 to 13% annual increases in their sewer and water rates? Something will have to give, and it will be operation and maintenance expenditures, as usual.

By the year 2010, the MWRA estimates its annual costs per household will approach $4000 — about 40 times the rates of 1980 (Figure 3.32). Something will have to change drastically. Modern Bostonians should not feel inadequate, however; they are simply making the same mistakes their ancestors made.

In addition, the $6 billion estimate for the Boston Harbor Project is recognized by serious planners to be inadequate to do the job, even temporarily. They realize that another several billion dollars will be needed to control and remove the raw sewage, pesticides, and urban filth that wash into the harbor every time it rains, to treat the highly contaminated rivers coming into the harbor, and to deal with the residual toxic materials accumulated throughout the harbor sediments.

The projections of sewer and water charges for the average household in the MWRA system indicated that it is unlikely that the MWRA would be able to properly operate the system, given its poor financial base. Thus, if the MWRA is forced by regulatory agencies to provide secondary treatment at Deer Island, it can be expected that the sewerage system will quickly deteriorate, and raw sewage will once again flow into the harbor and tributaries as pipes break and regulators malfunction.

This predicted failure of the newly constructed MWRA sewerage system will be the fourth nadir in the historical cycle of harbor contamination and remediation (Figure 3.30). A critical perusal of the history of Boston Harbor would have alerted planners to this recurring specter. Water planners and environmental engineers should study history.

Short- and Long-Term Solutions for Boston — If the MWRA were to shrewdly economize before its new system is completed — not by neglecting operation and maintenance of the plumbing, but by providing a minimal degree of treatment — it would be able to extend its honeymoon in the harbor significantly. This would take strong leadership and an appreciation for history, even though it would only be a short-term solution.

There may also be some ecological advantages to minimal or primary treatment instead of secondary treatment for the sewage discharged into the deep waters of Massachusetts Bay. This is especially likely when most of the toxic industrial wastes are removed from the MWRA sewerage system by its current programs. Effluent from a primary treatment plant has a more normal balance in the ratios of essential nutrients (carbon, nitrogen, phosphorus, and potassium) than does the effluent from secondary treatment plants. Thus, a primary effluent is more likely to stimulate normal biological productivity in Massachusetts Bay than would effluent from a secondary plant.

An ecologically balanced discharge without toxic substances would also be less likely to disturb the basic marine ecology and might even improve the food supply for whales migrating through Massachusetts Bay. The value of the discharge in increasing the krill populations fed upon by the whales should be investigated as a possible ecological justification for avoiding the expense of secondary treatment.

However, for the long range, the sole, really sustainable solution to protecting the ecology of Boston Harbor and Massachusetts Bay is to start decreasing the flows and waste loads by stabilizing the urban population, by resource and water conservation, and by generating revenue from the valuable organic wastes.

To do so will require engineers, planners, and political leaders to make some basic changes in their philosophies. They must go against the heady hubris of new public works projects such as dams, reservoirs, aqueducts, advanced treatment plants, and deep ocean tunnels. Instead, they must turn to the softer social programs of birth control, separation of waste streams, resource conservation, recycling, and other behavioral changes that can reduce consumption of water and production of sewage. Only by these measures can anyone expect to break the dismal cycle of Boston Harbor.

Citizens and thinkers of Boston have often led the nation in revolutions and new departures. There is still the capacity in New England to continue this behavior.

3.3.4 Merrimack River

The large, interstate Merrimack River has a shameful reputation as the source of the largest cholera epidemics of North America in the 19th Century, and for the complete elimination of the sturgeon and salmon populations from its waters due to the construction of dams and the poisoning of the river with industrial wastes. The Merrimack River drains a large part of New Hampshire, as well as the northeast portion of Massachusetts, and is a major source of fresh water coming into Massachusetts Bay.

Although there has been some success with restoration of salmon and shad in the Connecticut River, the Merrimack River is shackled by its industrial past. Restoration of these fisheries to the Merrimack River will take a major effort to modify the dams, contaminated canals, and leaking waste deposits along its banks.

The losses to New England of the fisheries resources of this once teeming river are significant. The huge sturgeon that used to ply the river were so big and so fast that it was dangerous to be out in the river when the sturgeon were running. These amazing fish live as long as people do, but reproduce very slowly. Thus, the primeval populations could not recuperate from the successive death blows of the industrial society that settled on the Merrimack River shores.

3.3.5 Small Coastal Rivers

In pleasant contrast to the Merrimack River, there are some coastal rivers on Massachusetts Bay with wholesome ecologies. Because of their small size and rich biological resources, some of these rivers have been maintained in fairly attractive condition, without severe changes in their primeval ecology. The Parker River, north of Boston Harbor, and the North River on the southern coast of Massachusetts are two examples of these clean rivers (Figure 3.33). The conditions and productivity in these rivers can be used to estimate what the productivity of the Merrimack River could be, and what we could expect in the tributaries of Boston Harbor if restored to primeval conditions. Thus, the small Parker River can play a large role as a model for the other rivers.

The few factories that discharged sewage into the Parker River were closed after World War II, and the subsequent recovery has restored the Parker River estuary to nearly primeval conditions. High levels of water quality documented in surveys in 1968 and 1969 met all of the state and federal standards for swimming and for marine fisheries (Table 3.4). Suspended solids and nutrients were typical of healthy conditions for tidal habitats.

There were about 600 acres of productive clam flats within the Parker River estuary, containing a standing crop of clams of 660 pounds per acre, including duck clams and blue mussels. With a standing crop biomass of 400,000 pounds of clams, the harvest during 1965 was 36,880 bushels of softshell clams.

Lobsters and crabs were harvested from the estuary in significant numbers. In 1965, about 10,000 lobsters were caught, and over the 2-year period from 1964 to 1965, about 3300 pounds of edible crabs were harvested from the estuary.

There were 27 species of fish in the estuary in 1965, the most abundant being winter and yellowtail flounder and the forage fish (mummichog, silversides, and stickleback). Striped bass were found throughout the estuary and in the surf off Plum Island from May to October. Mackerel were plentiful at the mouth of the estuary in summer, and white perch were caught in the upper estuary from spring to fall. In winter, ice fishing for smelt was popular.

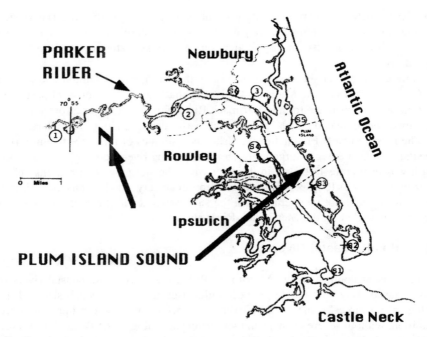

Figure 3.33   Plum Island Sound at the mouth of the Parker River. Sampling station locations for 1966 data shown in Table 3.4 are indicated by numbered circles.

Table 3.4   Measured Concentrations of Water Quality Parameters at Clean Water Stations on the Parker River near Byfield, 1966

Water quality parameter	Mean concentrations
Dissolved oxygen range, in ppm[a]	7.2–8.4
Suspended solids, in ppm	4
pH, in logarithmic units	7.6
Alkalinity, in ppm	37
Total coliform bacteria per 100 ml	300
Color, in standard units	68
Turbidity, in standard units	2
Ammonia as N, in ppm	0.04
Nitrites as N, in ppm	0.006
Nitrates as N, in ppm	0.1
Total phosphates as P, in ppm	0.16

[a] ppm, parts per million.

The vegetation of the Parker River estuary was typical of New England salt marshes, containing 22 species of plants dominated by the cord grasses, spike grass, seaside lavender, and glasswort. Nine species of algae occurred, with rockweed and sea lettuce most prevalent.

Chemical analyses of the flesh of clams and flounder in 1968 revealed small quantities of DDT at about 10% of the recommended maximum concentration, but there were no traces of other major pesticides, nor were there significant concentrations of heavy metals in these organisms.

3.4 RHODE ISLAND SOUND

Rhode Island Sound receives its major freshwater flows from Narragansett Bay on the southern coast of Massachusetts, which also drains the northeastern half of Rhode Island. The bay contains a beautiful complement of islands at its mouth, shielding numerous coves and harbors from hurricanes and North Atlantic storms. The annual Newport Jazz Festival each summer derives a great deal of its beauty from its setting near the mouth of the bay. Sailing and recreational use of the bay have supplemented the original fishing and shipping uses that developed over the last 2 centuries.

Coming from Massachusetts, the Taunton, Ten Mile, and Blackstone Rivers are the principal sources of flow into Narragansett Bay (Figure 3.34). These three small rivers have provided the process water for a host of small industries related to textiles, paper, and metal finishing since the 19th Century. Eli Whitney developed the cotton gin along the Blackstone River about 1870, giving the initial regional impetus to the textile industry.

These river basins presently contain 16 publicly owned plants for treatment of industrial and domestic wastewaters. The Fall River plant discharges to the bay, while the others discharge to freshwater rivers or the tidal reaches of the Taunton River.

3.4.1 Pawtuxet and Providence Rivers

Flow from Rhode Island reaches the bay from the Pawtuxet River and a score of coastal coves. In the drainage area feeding the bay, there were 19 treatment plants within Rhode Island in 1993, many of them with coastal discharges.

The two largest discharges were part of the Narragansett Bay Commission sewerage system, one entering the tidal portion of the Seekonk River at Bucklin Point, just south of Pawtucket. The flow from the Bucklin Point discharge was 15 cubic feet per second in 1993. The other discharge of the Narragansett Bay Commission entered the Providence River at Fields Point in Providence. This discharge was 30 cubic feet per second in 1993, twice the Bucklin Point discharge.

3.4.2 Taunton River

The Taunton River drains 530 square miles of southeastern Massachusetts, a broad, flat, and swampy basin. The river discharges to Mt. Hope Bay as it crosses the Rhode Island state line, flowing into the northeastern corner of Narragansett Bay before reaching the Atlantic Ocean south of Rhode Island. This section describes the dams and sewers along the mainstem of the Taunton River, along the Rumford River and Three Mile River tributaries, and those on Mt. Hope Bay (Figure 3.35).

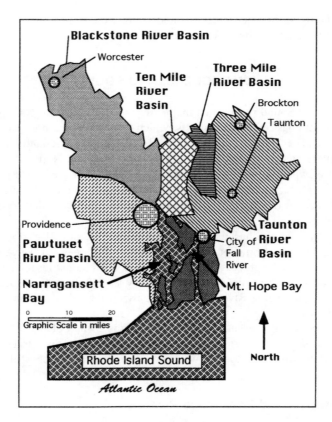

Figure 3.34   Rhode Island Sound Drainage Basin. The Narragansett and Mt. Hope Bay drainage basins cover 1820 square miles of Rhode Island and southeastern Massachusetts, including the islands and waters of the bays.

3.4.2.1 Fisheries

Numerous ponds in the Taunton River Basin contain native populations of bass
and pickerel. The river and tributaries once supported a very large alewife or herring
fishery and a moderately sized shad fishery. Since the colonial period, increases in
water pollution and construction of dams on many of the tributaries caused a steady
decline in these migratory fish populations. They were virtually eliminated by 1970,
when Massachusetts initiated a shad restoration project in the Taunton River.

On Earth Day 1996, there were small but significant signs of life in the Taunton
River and tributaries. The annual herring run filled the small basins and weirs of the
Nemasket River up to Quaker Mill with healthy adults, searching for their spawning
grounds. This occurred during a beautiful spring season that followed the snowiest
winter on record. Perhaps the melting snow and rushing torrents gave the herring
the chance they needed. Children gathered along the rivers, trying to grab the darting
schools of herring, laughing and shouting at their slippery elusiveness. The memory
of those rushing waters and jumping fish inspired some of these children to artistic

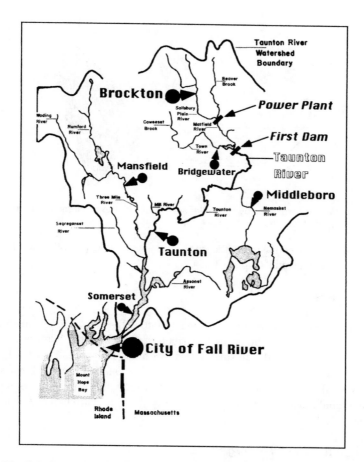

Figure 3.35 Sole dam on Taunton River, as well as seven sewage discharges in Mt. Hope Bay drainage basin, 1996.

creations (Figure 3.36). For a short, shining week, the primeval joy of Creation was real again. Would that it could happen every year!

In 1996, Mt. Hope Bay contained diminished populations of menhaden and many other species of marine fish. The Taunton River estuary and Mt. Hope Bay contained several species of shellfish, including softshell clams, the Eastern oyster, and the hard clam. A large potential for sustainable shellfishing exists in the estuary of the Taunton River, but nearly all shellfish areas were closed in 1996 to direct market harvesting, due to fecal contamination.

3.4.2.2 Hydrology

The mean annual flow from the Taunton River at its mouth is 430 cubic feet per second, including contributions from numerous tributaries and wastewater discharges (Figure 3.37). Travel times are long but extreme low-flow conditions are moderated

Figure 3.36 Artistic inspiration by child after seeing alewife run on Nemasket River, Earth Day
1996. (Drawing by Laura Jobin.)

somewhat in this river basin by the enormous swamps, especially the Hockomock
Swamp near Bridgewater (Figure 3.35).

3.4.2.3 Dams

The only dam on the mainstem of the Taunton River is 41 miles upstream from
the estuary (Figure 3.35). Some of the tributaries, such as the Rumford and Three
Mile Rivers, have over a dozen small dams, but many tributaries flow freely, making
possible the restoration of migratory fisheries. The Matfield, Mill, Nemasket, and
Town Rivers flow without serious physical obstructions from their sources to the
ocean, and could again support large migratory fish populations if the water quality
were restored.

3.4.2.4 Industrial Discharges

A major change in the distribution of wastewater discharges in the last three
decades has been the elimination of individual industrial discharges, primarily by
connecting them to expanded municipal sewer systems. In 1970, there were over 40
individual industrial or municipal discharges, plus innumerable combined sewer
overflows. However, by 1996 most of the industrial discharges had been connected
to the five municipal systems that discharged to rivers. Also, the combined sewer
overflows had been improved, thereby reducing the extent of overflows.

This diversion of flow, including further diversions from domestic sewers that
collect water pumped from upstream aquifers and then discharge it downstream of
the towns, has reduced the flow and assimilative capacity of most of the peripheral
streams. The collected wastewaters are then discharged at a single point for each
municipality or region, turning the river downstream into an open sewer during
summer.

Although these regional systems are usually more economical to construct than
individual treatment facilities, their impact on hydrology and aquatic ecology can
be devastating, especially in the upstream reaches of small rivers.

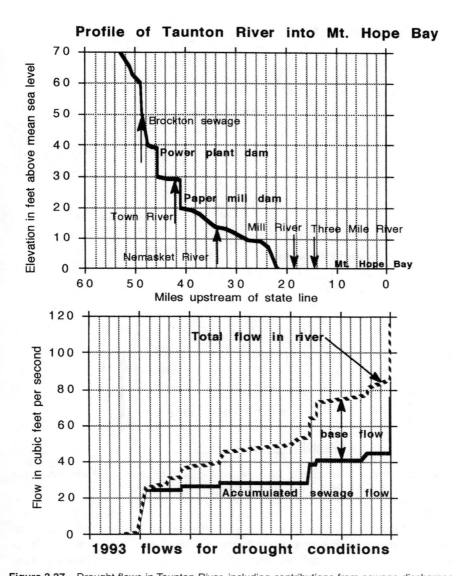

Figure 3.37 Drought flows in Taunton River, including contributions from sewage discharges.

3.4.2.5 Combined Sewer Overflows

The city of Fall River is the largest and most industrialized community on Mt. Hope Bay. A substantial pollution load was delivered to Mt. Hope Bay during the 1970s from Fall River's combined sewer overflows, even during dry weather. The old combined sewer system of Fall River was a legacy of the mistaken design concepts applied along the entire east coast of North America as the colonial settlements developed into large cities.

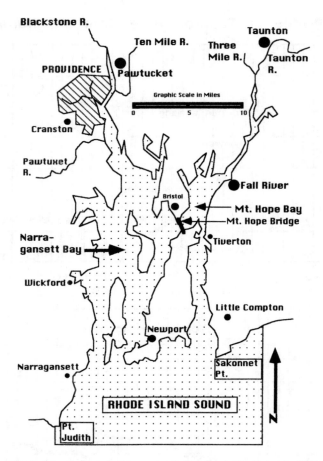

Figure 3.38   Rhode Island Sound, Narragansett Bay, and tributaries in Rhode Island and Massachusetts.

During heavy rainfall, there were 24 combined sewer structures discharging a mixture of untreated sewage, rainwater, and urban runoff. The pollutant loads from these overflowing sewers were about the same magnitude as that of the Fall River treatment plant, and their diffuse nature resulted in widespread fecal contamination of the river and bay, as measured in concentrations of fecal coliform bacteria.

A computer simulation indicated that contaminated water would move toward Mt. Hope Bay along the deep Taunton River channel and become trapped near the mouth of the Tiverton channel (Figure 3.38). After a few days, it was gradually dissipated by tidal exchange under the Mt. Hope Bridge.

3.4.2.6 Water Temperature

Surveys in the summer of 1970 indicated that water temperature rose to very high values at the mouth of the Taunton River. Although the upstream temperature

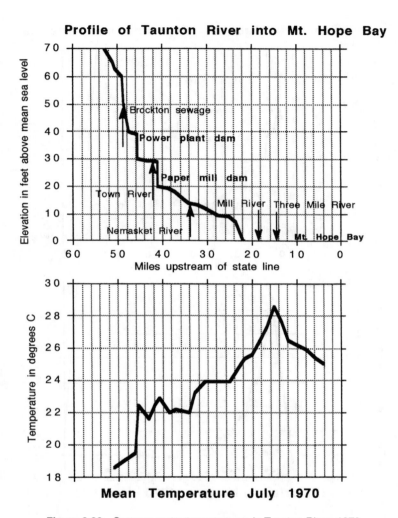

Figure 3.39 Summer water temperatures in Taunton River, 1970.

near Brockton was less than 19°C, the temperature rose to nearly 29°C at the confluence with the Three Mile River (Figure 3.39). Thus, the entire tidal portion of the Taunton River had temperatures so elevated they would cause serious problems for fish. During this same survey, dissolved oxygen concentrations were nearly zero. A large portion of this thermal pollution came from large power plants on the shores of Mt. Hope Bay that were using the water for cooling.

3.4.2.7 Nutrients

In 1970, the concentrations of phosphate nutrients in the upper Taunton basin, namely the Salisbury Plain River of Brockton, exceeded 15 parts per million (ppm) due to industrial discharges, as well as contributions from domestic sewage

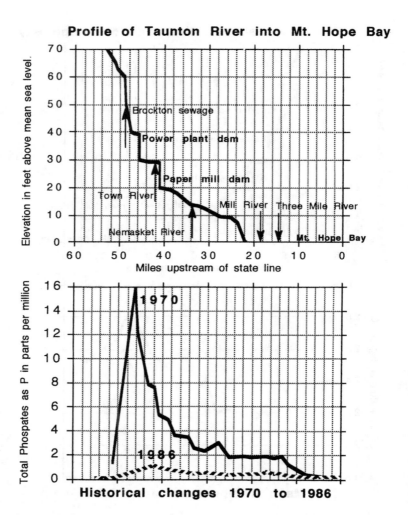

Figure 3.40 Phosphate nutrients in Taunton River, 1970–1986.

(Figure 3.40). This exorbitant concentration occurred downstream of the Brockton sewage discharge. Unusually high concentrations of phosphates were found all the way to Mt. Hope Bay that year.

By 1986, after addition of extensive phosphate removal systems at most of the municipal plants, the peak phosphate concentration was reduced to 1 ppm, and the downstream values were reduced to 0.2 ppm (Figure 3.40). Despite this 94% reduction in phosphates, the residual concentrations in the river caused excessive algae growths, even in 1986.

The large algae populations caused wide fluctuations in dissolved oxygen, reaching low concentrations on hot, cloudy days, to the detriment of the fish populations and other users of the rivers.

I will never forget my first sampling run in 1970 on the Taunton River system, working for the Division of Water Pollution Control. One of my sampling stations was on the Salisbury Plain River just below the discharge of the old Brockton Sewage Treatment Plant. For the photosynthesis runs, I had to suspend racks of bottles from floats in the river. One set was just below the surface; another was deep at the Secchi Disk depth; and a third was suspended halfway between the first two.

On the three hot sunny days of that summer survey, I had to roll up my sleeves and reach down to grab the lower set of bottles. When my arm came out of the river, it was coated sick-green with algae, beautifully dotted with red sludge worms. Red and green like a Christmas tree.

I remember the stench, too. I could tell when I was getting near the river by the ripe, sewage smell. It reminded me of a malfunctioning sludge digestor.

That winter, we brought the Town of Brockton and their consulting engineer into the Boston offices for enforcement proceedings. The engineer blithely told us that the secondary treatment plant was one of the most modern in the state, and that one could almost drink the effluent! That is what he had been taught in engineering school. I gave him a first-hand description of what his marvelous treatment plant was really doing to the Salisbury Plain River. I remember wishing he had looked at the river — just once.

3.4.2.8 Dissolved Oxygen

An early water quality survey during the Great Depression and before World War II determined that the organic pollution of the Taunton River was severe, due to discharges of the numerous industries in the basin, as well as untreated municipal wastewaters. A measure of this organic pollution load is the 5-day Biochemical Oxygen Demand (BOD5), which should not go above 1 part per million (ppm) in clean water. However, it repeatedly exceeded 4 ppm in the Taunton River in 1938. Despite periodic programs for control of these discharges, the BOD values in 1986 were only slightly lower. Downstream of the new discharge of sewage from Bridgewater, the BOD rose to 6.5 ppm, higher than any concentrations measured half a century earlier.

Primary sources of this organic material in 1986 were sewage plants serving Brockton, Bridgewater, Middleboro, Taunton, and Fall River. In 1938, the amount of wastewater discharges recorded in the basin was 1.5 cubic feet per second (cfs), rising to 75 cfs by 1993. The slow improvements in treatment facilities hardly kept pace with population growth.

In 1996, there were only seven treated sewage discharges in the Taunton River Basin, the largest from Fall River, discharging directly into Mt. Hope Bay (Table 3.5). The total discharge from these plants in 1996 was 110 cfs. A careful analysis of the receiving waters was made for each discharge, resulting in specification of maximum limitations on certain chemical parameters. These maximum limits were necessary in order to meet the state water quality standards for the rivers and bays. In the case of dissolved oxygen, a minimum limit was established for the discharge, sometimes requiring oxygenation of the plant flow.

Table 3.5   Municipal Discharges in Taunton River System and Mt. Hope Bay, 1996

Town	Flow, in cubic feet per second	Biochemical oxygen demand (BOD5) limit in effluent, in ppm	Ammonia limit in effluent, in ppm	Phosphorus limit in effluent, in ppm	Oxygen required in effluent, in ppm
Bridgewater	10				
Brockton	28	5	1	1	7
Fall River	48	30			
Mansfield	5	10	1	1	6
Middleboro	3	7	1	1	7
Somerset	3	30			
Taunton	13	15	1		
Totals	110				

Note: Required limits were developed from an analysis of the state water quality standards for the receiving waterbodies.

As this organic material decomposes in the water, the biochemical and bacterial processes consume oxygen, causing the oxygen to decrease to values that kill fish. In the summers of 1970 and 1975, dissolved oxygen concentrations throughout the Taunton River fell below the 5 ppm needed for warm-water fisheries (Figure 3.41). Improved sewerage systems and new treatment facilities reduced the organic loadings by 1986, giving the river a chance to recover somewhat.

In 1986, however, the oxygen concentrations still fell below the desirable value in the upper reaches of the river, for over 10 miles below the Bridgewater sewage discharge, and for at least 5 miles below the Taunton sewage discharge. Large daily fluctuations in oxygen were due to excessive algae populations fertilized by the domestic sewage. Fluctuations in the estuary and Mt. Hope Bay often exceeded acceptable ranges, and created dangerous and unstable conditions for fish populations.

3.4.2.9   Solids

Suspended solids in the Taunton River have been very high over the last two decades, decreasing slightly as sewage treatment facilities were improved (Figure 3.42). In 1970, the suspended solids exceeded 50 ppm downstream of the Bridgewater sewage discharge and in Mt. Hope Bay downstream of the Taunton sewage discharge. Clean water has suspended solids concentrations below 5 ppm.

By 1986, the increased treatment had improved the conditions somewhat, but concentrations of 30 ppm occurred below the Bridgewater sewage discharge, and values in Mt. Hope Bay often exceeded 10 ppm.

3.4.2.10   Herring Runs on the Taunton River, 1996

Although some herring (alewife) runs continue fairly strongly in short, coastal rivers in New England, the Taunton River is one of the few large rivers in New England that continues to support extensive herring migration. The first dam across the main course of the Taunton River is 41 miles upstream of the river mouth,

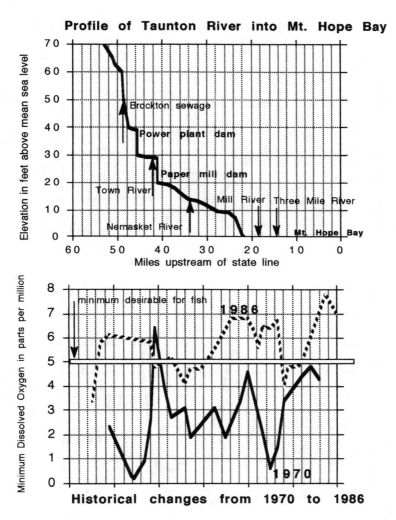

Figure 3.41 Minimum dissolved oxygen concentrations observed in Taunton River, 1970–1986.

providing a large expanse of river and tributaries for herring habitat. The most prolific herring run is up the Nemasket River tributary, which flows through Middleboro from a series of lakes. Only a few low sills and colonial-era weirs cross the Nemasket tributary, allowing herring to return to the headwaters every year. The estimated number running in April and May of 1996 was 1 million herring.

The magnitude of these herring or alewife runs was impressive, especially in the face of the chemical and thermal pollution of Mt. Hope Bay, the Taunton River estuary, and the urbanized section of the Taunton River, which these small migrants must pass. Their continued existence gives hope that such runs might be restored in the rest of coastal New England.

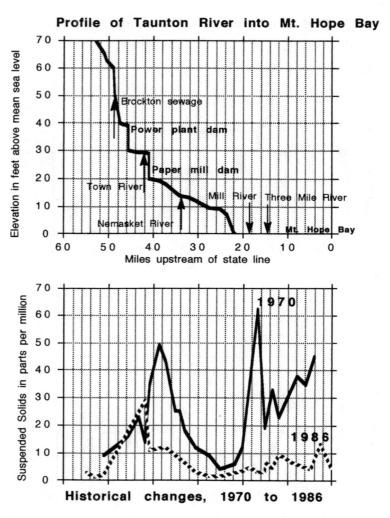

Figure 3.42 Suspended solids concentrations in Taunton River, 1970–1986.

Other well-documented herring runs in Massachusetts during 1996 included annual runs of 350,000 in Bourne on the Cape Cod Canal and 600,000 in Weymouth Back River. Many towns employ part-time herring officers who prevent poachers from scooping up the dense swarms of struggling fish and also maintain the small weirs and sills that assist the herring in their progress. During the Great Depression and shortly after World War II, these herring officers were fully occupied as the herring were an important seasonal food source, very much appreciated by low-income families. During that period, fishing in general was an important part of the New England economy, before neglect of aquatic and marine resources led to its collapse.

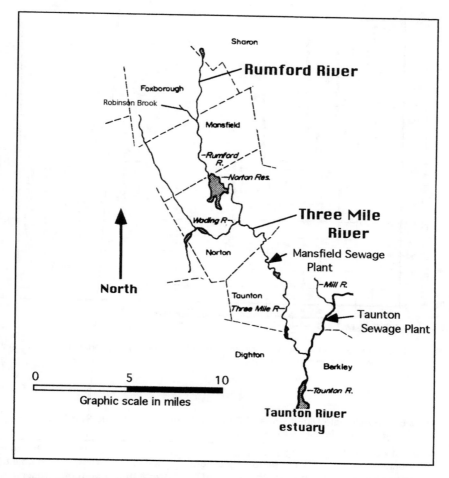

Figure 3.43 Rumford and Three Mile Rivers in Narragansett Bay drainage area.

3.4.3 Three Mile River

On the mainstem of the Three Mile River, there were 12 small dams in 1996 (Figure 3.43). The dam at Norton Reservoir was the highest (Figure 3.44). Migrating fish would have to jump 12 feet vertically to pass it (Table 3.6). The Norton Reservoir also had the largest storage volume — 70 million cubic feet when full to spillway elevation. The small dams and reservoirs have had detrimental effects on the ecology and fisheries of the Three Mile River, in a variety of ways illustrated in the following sections.

For a typical and robust springtime flow of 290 cubic feet per second (cfs) at the confluence with the Taunton River, the time of travel from the inlet of Norton Reservoir to the Taunton River is 110 hours. The usual critical flow for water quality

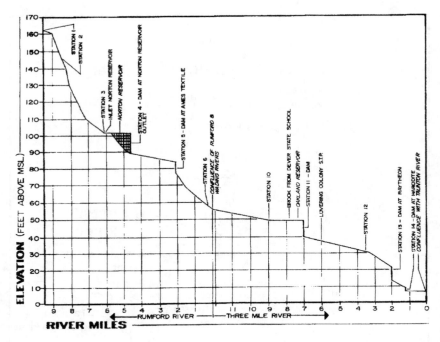

Figure 3.44 Location of five major dams on Three Mile River. An additional 10 smaller dams are found on the tributaries.

Table 3.6   Approximate Dimensions of Dams and Reservoirs on the Rumford and Three Mile Rivers

Name of dam and distance from Taunton River confluence, in miles	Height for fish to jump in order to pass over dam, in feet	Length of reservoir, in feet	Mean width of reservoir, in feet	Mean depth of reservoir, in feet	Storage volume of reservoir, in million cubic feet
Morse Brothers 25	3	2000	300	2	1.2
Gavins 23	6	1600	600	4	3.8
Vandys 22.5	5	1800	200	2	0.7
Bleachery 21	6	900	200	2	0.4
Fulton 18	5	1000	400	3	1.2
Kingman 17	4	800	300	2	0.5
Cabot 16	3	800	200	2	0.3
Norton 15	12	8000	4000	3.2	70
Read and Barton 14	6	1000	500	3	1.5
Oakland 7	10	1000	800	5	4
Raytheon at Dighton 2	10	500	200	5	0.5
Harodite 1.2	3	500	200	2	0.2
Total storage, in million cubic feet					84.3

criteria determinations by state and federal regulatory agencies is the lowest flow over 7 consecutive days, with a return period of 10 years. Analysis of historical data determined that this critical flow is 7.4 cfs at the mouth of the Three Mile River. This 7-day, 10-year low flow would require an interminable travel time of 4500 hours or 190 days from the inlet of the reservoir to the mouth of the Three Mile River during such drought conditions. These travel times were estimated from measurements of dye patches in 1969 and 1970.

Because of the considerable storage provided by the existing 12 dams in the Three Mile River Basin, current floods are significantly moderated in comparison to floods that occurred in the primeval state before any dams were built. During spring storms, the reservoirs are usually full, so the volume of spring floods is not decreased by the dams. However, the length of the flood is extended as it passes through the reservoirs, consequently reducing the peak discharge, even without changing the total volume of the flood.

3.4.3.1 Trapping of First Flush of Floods by Dams

During hurricanes that occur in the autumn when the reservoirs are usually empty, the dams reduce the volume of the flood by trapping the first part of it (Figure 3.45). Based on the 84 million cubic feet of storage available in the 12 dams and the current runoff coefficient for the basin of 0.57, the dams would trap all hurricanes or autumn storms with rainfalls of 1.75 inches or less. For larger storms, the flushing effect would also be proportionally reduced.

This combined impact of reduction and dispersion of floods solely due to the reservoir volumes has markedly reduced the annual flushing of debris down this river into the Taunton River and Narragansett Bay. The wood and organic portions of this debris are an important source of organic carbon that provides the base of the food chain for aquatic and marine organisms. Loss of this carbon transport causes a reduction in the overall productivity of the estuary and coastal waters, compared to primeval conditions.

3.4.3.2 Change in Runoff due to Watershed Modifications

Human population growth and consequent modification of the watershed has shortened the travel time of rainfall in the basin, due to changes in land use and surface characteristics of the land. This modification compensates somewhat for the longer travel time caused by the dams. Human activities in the watershed have shortened the travel time by causing an increase in the amount and velocity of rainwater flowing overland toward the stream channels. This is due to loss of forest cover, and the loss of absorbing capacity of the land due to pavement, houses, and parking lots. Also, a large proportion of the primeval wetlands have been filled, eliminating natural storage, evaporation, and infiltration of rainwater. Storm sewers designed to avoid flooding in populated areas also hasten the flow of rainwater to stream channels.

Figure 3.45   Raytheon Dam in Dighton. The dozen dams similar to this low one have relatively little storage volume at spillway elevation, but can cause considerable bank storage when a large flood passes over them.

These modifications also decreased the amount of rain that is absorbed into the vegetation and soil, decreased the flow into the underground aquifer and thus the base summer flow in the rivers, and changed the character of the suspended and dissolved material washed into the rivers with the spring rains and autumn hurricanes.

In terms of storm hydrology, the effect of this watershed modification was to shorten the travel time for overland travel of water, and to increase the amount of this flow. From maps and measurements of the watershed character for 1975, this change was estimated to shorten the travel time in the watershed around Lake Wolomolapoag at the headwaters by 50%. Thus, the current overland travel time to the lake is 1 hour, while the primeval travel time was about 2 hours. The runoff coefficient for the entire basin was assumed to be 0.40 for the primeval forest, and 0.57 for the current situation. The increased runoff coefficient due to watershed modification was estimated from land patterns measured in 1975. The primeval distribution of forest, wetlands, and impervious area has been drastically altered (Table 3.7).

Flow models developed for hurricane conditions indicated that hurricane flood waves are considerably delayed and moderated by the dams, although there is some hydraulic compensation due to human modifications.

Removal of all the dams would decrease the current time of travel for the entire Three Mile River Basin from 9 to 2.3 hours, for hurricanes with 100-year return periods, if the primeval watershed conditions were also restored. With the shorter time of travel, higher-intensity rainfalls would be experienced over the entire basin and would increase the peak flow from the current 14,000 cfs to a primeval peak

Table 3.7   Changes in Land Use of Three Mile River Watershed

Land use	Primeval condition, percentage	Norton Reservoir watershed 1975, percentage	Estimated Three Mile River Basin, square miles
Forest	98%	64%	54.3
Open land	1%	7%	5.5
Open water	1%	4%	3.5
Residential		19%	16
Agriculture		4%	3.0
Industry		2%	1.5
Total	100%	100%	83.8

Note: Current conditions for basin were taken from map measurements for watershed above Norton Reservoir in 1975. Primeval conditions were estimated based on descriptions of early European immigrants.

Table 3.8   Delay in Hurricane Peak Discharges due to Small Dams on the Three Mile River

Character	Units	Primeval condition without dams	Current condition in 1996 with 12 dams
Travel time	Hours	2.2	9
Rainfall intensity	Inches per hour	1.30	0.47
Drainage area	Acres	53,600	53,600
Runoff coefficient	None	0.40	0.57
Peak flow when entire basin is contributing	Cubic feet per second	28,000 (47,000 with current land use)	14,000

Note: These conditions were calculated for flows at the mouth of the river due to hurricanes with 100-year return periods, using the Rational Method.

flow of 28,000 cfs, due to the shortened time of travel through the watershed (Table 3.8).

With the present land use and watershed modifications, removal of all the dams would decrease the travel time to only 1.23 hours, with a drastically increased peak flood at the river mouth of 47,000 cfs, thereby exceeding the primeval flood of 28,000 cfs. A better simulation of primeval conditions would be to remove only some of the dams, striving for a travel time equal to the primeval travel time of 2.2 hours.

3.4.3.3   Water Temperature

Using a simple hydraulic model, the impact of the dams on water temperature was estimated, after calibrating the model with temperature data from field surveys in 1988. The estimated temperature difference between the present condition and the primeval condition was considerable, with a 2°C lower temperature predicted at the river mouth for the primeval condition without dams (Figure 3.46). The major increase in temperature occurred as the water flowed through the shallow Norton Reservoir, where it was exposed to direct solar radiation with no shade.

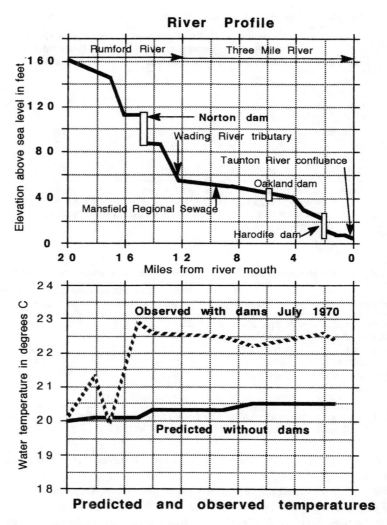

Figure 3.46 Predicted and observed water temperatures in the Three Mile River. A simple
calibrated model was used to predict the river temperature in the absence of
dams, freely flowing, and then compared with the observed temperatures in 1970.

The model was calibrated with two adjustments for local conditions. Upstream
of Norton Reservoir where the river was shallow and flowed through open and fairly
urban areas, the fitted rate of temperature increase for the model was 0.01°C per
hour. From Norton Reservoir downstream where the river was deeper and flowed
through wooded and shaded fields, the temperature increased at 0.002°C per hour.
Shading effects from riparian trees along the narrow reaches of the river were also
included. The impact of temperature in Norton Reservoir as a function of seasonal
and flow conditions is further explored in a subsequent section of this chapter.

3.4.3.4   Water Quality

Detailed water quality surveys on the Three Mile River since 1970 have shown several improvements in the quality of the river, primarily due to the construction of a regional sewage plant downstream of Mansfield in 1986 (Figure 3.43). The combined flows from Foxboro, Mansfield, and Norton were diverted to this location and given tertiary treatment including phosphate removal, aeration, and chlorination of the effluent. The impact of this new treatment plant was evident in the decreases in suspended solids and total phosphate concentrations in the river, and to a lesser extent in the improvements in dissolved oxygen.

Human watershed modification also had an impact on water quality due to a change in the type of material washed into the rivers. This material changed from topsoil, vegetation, and forest debris to lawn fertilizers and pesticides, road salts, and particulate matter from roadways. The waste material washed off roadways is mostly grease, oil, rubber, and heavy metals (including lead from fuel additives before 1990).

Suspended Solids — Suspended solids in the river exceeded 22 parts per million (ppm) in Norton Reservoir during the late summer of 1970 while Mansfield was discharging sewage into the reservoir (Figure 3.47). This high concentration decreased only slowly downstream and never dropped below 7 ppm. However, in the summer of 1988, the lower reaches of the river had only about 4 ppm of suspended solids. The magnitude of the decrease is greater than indicated by these values because the 1988 river discharge at the mouth was only 12 cubic feet per second (cfs), compared to 40 cfs during 1970. Thus, the load of suspended solids in the river in 1988 was only 11% of the load measured in 1970.

In August 1991, the impact of a hurricane on suspended solids was measured in the Rumford River and through Norton Reservoir. Although concentrations in the river upstream of the reservoir had only been 2 ppm before the hurricane, they rose to 10 ppm by the time the rain stopped. This increase occurred despite several small dams upstream. Suspended solids in primeval rivers in Massachusetts were probably below 1 ppm, except after storms. This is the concentration in headwaters of pristine river basins in Massachusetts at present. The hurricane caused a drastic increase in suspended solids in the reservoir, reaching 26 ppm shortly after the storm (Figure 3.47).

Total Solids — Another simple indicator of water quality is total solids, which have been measured in Massachusetts rivers as far back as the 19th Century and are often summarized as total residue. This parameter indicates the amount of material remaining after a liter of river water is boiled away. Total solids thus includes the sum of suspended solids and dissolved solids, and usually does not exceed 50 ppm during dry weather in the absence of discharges of human or industrial wastes.

Measurements in 1970 indicated total solids of 170 and 140 ppm in the Rumford River and the Three Mile River, respectively (Table 3.9). During a more recent survey in 1988, the total solids were 160 and 140, respectively, remaining significantly above the estimated primeval concentration of 50 ppm. Much of this excessive total solids in 1988 was probably due to chemicals applied to lawns, transportation residues on roads and highways, and from contamination of groundwater with domestic sewage from individual homes.

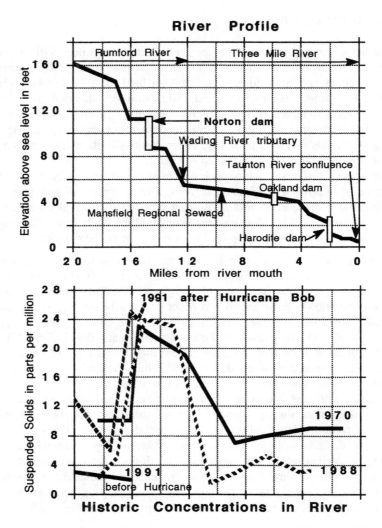

Figure 3.47 Suspended solids in Rumford and Three Mile Rivers. The decrease in concentration from 1970 to 1988 was due to a new sewage treatment plant constructed downstream of Norton Reservoir in 1986. However, a hurricane in August 1991 increased the concentration of suspended solids markedly, where measured in Norton Reservoir and upstream.

Table 3.9   Total Solids in Rumford and Three Mile Rivers

Year	Mean for Rumford River, ppm	Mean for Three Mile River, ppm
1988	160	140
1970	170	140
Primeval estimate	50	50

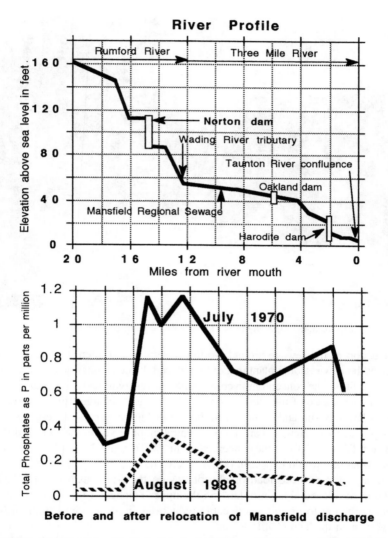

Figure 3.48   Phosphorus nutrients in Rumford and Three Mile Rivers, 1970–1988. The relocation of the Mansfield sewage discharge from Norton Reservoir to a location downstream of the Wading River confluence, as well as the increased degree of treatment provided at the new plant, caused almost a 90% decrease in the concentration of phosphates in the river by 1988.

Nutrients — A major cause of the weed and algae problems in Norton Reservoir and the Three Mile River during the 1970s was the excess of phosphate nutrients coming into Norton Reservoir in the Mansfield sewage discharge. The accumulated concentrations in the reservoir exceeded 1 ppm during the summer of 1970, and remained above 0.6 ppm downstream (Figure 3.48). These concentrations were much too high for a normal river.

Figure 3.49   Spillway of Norton Reservoir Dam on Rumford River.

Diversion of the sewage around Norton Reservoir and intensive treatment of the sewage at the new regional plant markedly reduced the concentration of phosphates by 1988. In addition, the concentration coming into the reservoir had been reduced to clean water values because of the elimination of the sewage discharge from Foxboro in the Rumford River Basin, and reduction in the use of high-phosphate detergents in Massachusetts due to a statewide ban. The concentration observed in the Rumford River in 1988 was 0.03 ppm, 10% of the concentration in 1970 (Figure 3.48). This concentration is about what one would expect in primeval streams.

Although the reduced load of phosphates on the river basin was marked, and resulted in decreased algae problems in the upper reaches of the Rumford River, it was not sufficient to produce noticeable reductions in algae populations in Norton Reservoir or in the Three Mile River.

To control algae growth, phosphates in the river must be reduced below 0.05 ppm. The sediments of Norton Reservoir contained high concentrations of phosphates that came back into solution under certain conditions, thus causing a rise from 0.03 ppm at the inlet of the reservoir to 0.38 ppm at the outlet (Figure 3.48). Continuation of the severe algae and weed problems motivated the people of Norton to buy the dam and reservoir from its industrial owners in 1990 in order to dredge the sediments (Figure 3.49). It was hoped that this would restore the reservoir to its normally attractive ecology.

3.4.3.5 *The Heat Death of Norton Reservoir*

Limnologists, like many scientists, often ignore the most basic aspects of their science while concentrating on the most exotic. In the case of Norton Reservoir and

When Norton Reservoir filled with sludge and nutrients from the discharge by the Mansfield sewage treatment plant, I was surprised the remedy proposed to the Town of Norton was to:

1. Build a new regional treatment plant downstream on the Three Mile River near Taunton, and divert all the sewage flow around Norton Reservoir;
2. Purchase the reservoir and downstream water rights so the reservoir would no longer be emptied in the late summer and would thus be available for recreational use;
3. Dredge Norton Reservoir to make it deeper and thus reduce emergent weed growth.

Unfortunately, these solutions ignore the basic flow and temperature problems that plague Norton Reservoir.

many other reservoirs in New England, the neglected aspects are flow and temperature. These have had more to do with the death of these reservoirs than have the exotic algae and contaminants usually monitored when the reservoirs are studied.

The annual drawdown of Norton Reservoir to feed industrial uses downstream was disruptive for recreational uses, but was probably a major factor in maintaining the suitability of the reservoir for fish. If the reservoir was not drained in late summer, it would have become too hot, thus developing enormous algae populations and recycling nutrients that would otherwise be flushed out each autumn.

In July 1970, the flow coming into the reservoir was 8 cubic feet per second (cfs) from the Rumford River and 1 cfs from the Mansfield treatment plant on Back Bay Brook (Figure 3.50). The volume of the reservoir at spillway elevation was about 22 million cubic feet; thus, the theoretical detention time for the flow through the reservoir was 28.5 days. The temperature rose from 19.9 to 22.9°C at this low flow, a rise of 3° (Figure 3.46).

Impact of Flow Diversion on Temperature — Diversion of the Mansfield sewage flow around the reservoir by relocating the sewage treatment plant downstream in 1986 also reduced the flow into the reservoir, causing additional temperature increases in the summer due to the longer residence time of the flow through the reservoir. The summer flow was further reduced because the new Mansfield plant also accepted sewage flow from Foxboro, which had previously discharged to Robinson Brook, and from three other small sources in the watershed (Figure 3.43). In addition, the homes on the shores of Norton Reservoir were connected to a perimeter sewer that prevented their septic effluents from reaching the reservoir.

The flow budget estimated for a drought year indicated that the net effect of diverting the Mansfield sewage flow around Norton Reservoir would cause the flow out of the reservoir to cease during the month of July, and cause a slight decrease in the reservoir level (Figure 3.51). The theoretical detention time in such a case, calculated on the basis of the flows coming into the reservoir, would increase from 40 to 100 days.

Temperature increases in the reservoir correlated negatively with the flow rate into the reservoir, as did the temperatures of the incoming flow (Figure 3.52). Combined together, these correlations indicated very high temperatures in July for drought conditions. The water temperature in a drought year would increase 5.4°C, rising to 30.4°C at the dam. Without the diversions, the temperature at the dam

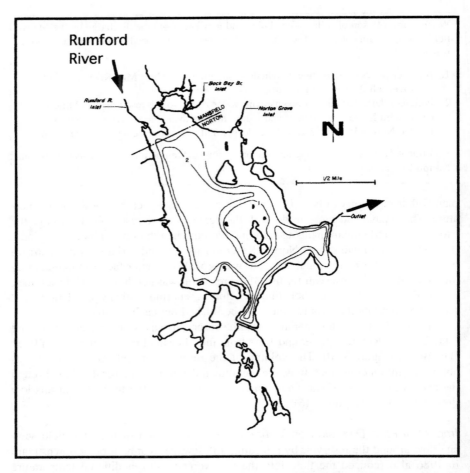

Figure 3.50   Norton Reservoir on Three Mile River. The contour lines indicate mean depth of
water in meters. (Reproduced with the permission of the Conservation Commission, Norton, Massachusetts.)

would have been much lower, about 23°C. This is a serious temperature distortion
that has not been recognized in planning for rehabilitation of the reservoir ecology.

The increase in temperature at the dam would be due both to solar heating during
slower travel down the Rumford River and to the longer detention time in Norton
Reservoir. A surface water temperature of 31°C was measured near the dam in July
1981 when the flow into the reservoir was very low, at 3 cfs, indicating that these
predictions are reasonable.

3.4.4   Mt. Hope Bay

There were two important conclusions to derive from an analysis of the data on
aquatic ecology from Mt. Hope Bay and the rivers that feed it. The first concerns

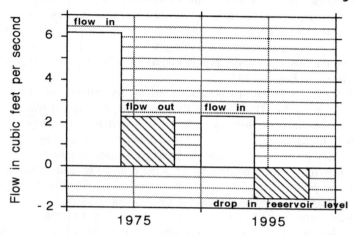

Figure 3.51   Flow budgets for Norton Reservoir during drought years. In 1975, the sewage discharges from Foxboro and Mansfield flowed into the reservoir, but they had all been diverted below the reservoir by 1995.

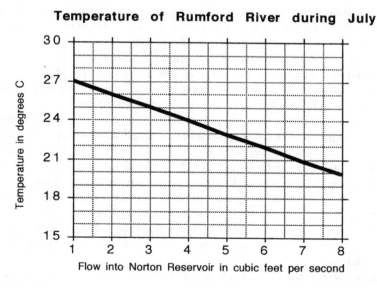

Figure 3.52   Flow rate of Rumford River coming into Norton Reservoir and temperature in the reservoir.

the resource value of suspended solids from the sewage treatment plants as they enter the bay. The second concerns the need for reductions in the expensive summertime treatment required for large sewage plants that discharge into small tributaries.

3.4.4.1 Fisheries Productivity of Mt. Hope Bay

Sewage can be seen as a waste causing degradation of the environment and offensive conditions, or conversely it can be seen as a source of nutrients and food for small aquatic and marine organisms. If it does not contain toxic industrial chemicals such as heavy metals and biocides, its food value at the base of the food chain can stimulate fisheries under proper conditions. This was the situation documented in Mt. Hope Bay in 1983.

By studying the relationships between nutrients and aquatic organisms in the Taunton River and in Mt. Hope Bay at the mouth of the river, it was found that moderate increases in river flow and nutrient transport into Mt. Hope Bay were correlated with increased productivity of algae populations in the bay, and that this increased primary productivity resulted in greater zooplankton and fish larvae populations. Furthermore, the populations of winter flounder and anchovy larvae increased as the productivity and zooplankton increased. Thus, increased sewage discharged to the bay caused an increase in the fish populations.

There were some limits to this productive relationship. Extremely high spring floods in the Taunton River that occurred during the period of spawning reduced larval fish densities. This was probably due to hydraulic washout of the larvae and their food, with rapid transport of the nutrients to the ocean. However, for moderate flow rates, there was a positive and linear relation between river flow and larval abundance.

The relation of nutrients and productivity had a seasonal complexity. In the summer, nitrogen limited the productivity of algae in Mt. Hope Bay, whereas in the winter, organic carbon was limiting. Low flow of the Taunton River at the end of summer was associated with decreases in nitrogen and other lesser nutrients, thus causing decreases in productivity in the bay. The long travel time of the river during low-flow conditions resulted in almost complete degradation of the organic material coming from sewage discharges, by the time it reached the bay.

However, when flows were high, especially in the spring or early summer, the nitrogen concentration in the river increased dramatically because there was not enough time for complete degradation. This increase in load of nitrogen on the bay resulted in increased productivity and increased fish populations. Similarly, moderate spring floods increased nutrient loads coming from agricultural runoff and even from overloaded sewage plants. It is important to note that the several treatment plants on the Taunton River system have low amounts of heavy metals and other toxic contaminants. Thus, the organic carbon, nitrogen, and phosphorus function as nutrients to increase the supply of fish food without being inhibited by toxic materials.

Such conditions indicated the resource value of sewage for fisheries productivity, as long as it is free of industrial wastes and other toxic materials. With proper treatment and placement of the treated discharges, domestic sewage treatment plants without industrial contaminants can thus become sources of increased fish populations, rather than causes of fish kills.

Table 3.10   Sewage Treatment Facilities in the Narragansett Bay Drainage Basin Required to Give Extra Treatment During the Summer

Town	Year of plant construction	Discharge, in cfs	Receiving water body	Low flow of river, in cfs
Brockton	1984	28	Salisbury Plain River	1
Taunton	1978	13	Taunton River	40
Mansfield	1985	5	Three Mile River	4
Attleboro	1980	13	Ten Mile River	6
North Attleboro	1980	7	Ten Mile River	0.5

3.4.4.2   Late Summer Operation of Treatment Plants

Despite extensive efforts and programs since Earth Day, the suitability for fish populations of the rivers and tributaries leading to Mt. Hope Bay has barely improved. The improvements in dissolved oxygen and nutrients in the freshwater portions of the Taunton River were achieved at exorbitant expense, requiring advanced waste treatment for all the major municipal systems. In summer, when oxygen values decrease the most, meeting the standard of 5 ppm for fish becomes increasingly expensive. Most of the major towns in the basin must increase the degree of treatment of their wastes during the months of June, July, and August (Table 3.10).

Despite advanced treatment, the low flow in the rivers does not provide enough dilution for nutrients; thus, algae populations overpopulate the impoundments downstream of the treatment plants and oxygen concentrations fluctuate wildly each summer day.

One sensible solution to the problem of overfertilization of the rivers during the summer is to use the treated effluent to irrigate crops along the banks of the rivers. Such a disposal system could be operated largely by gravity, and would save considerable operating expenses in the summer because the effluent would not require such a high degree of treatment.

Each of these sewer authorities should search for suitable land downstream of its discharge points, to evaluate its use for irrigating alfalfa, orchards, ornamental flowers, or other crops not posing a health risk.

The annual reduction in operating and chemical costs for the treatment plants in Norton, Taunton, Fall River, Middleboro, Bridgewater, and Brockton should save several million dollars each summer. To this should be added the profit from the farming. Farmers should be happy to receive the benefit of a steady stream of fertilized water during the normally dry summer months.

Another sensible solution to excessive algae populations in the Taunton River system would be to eliminate the old dams, thus shortening travel times in the river and lowering the effects of solar heating. The resultant cooler water will absorb oxygen more rapidly and thus favor fish populations as well.

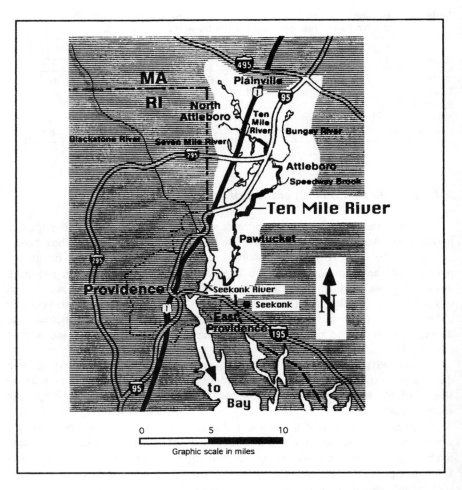

Figure 3.53   Ten Mile River near Providence, Rhode Island. Light area indicates watershed contributing flow to river.

3.4.5   Ten Mile River

The Ten Mile River and its two main tributaries — the Seven Mile and Bungay Rivers — drain 50 square miles in southeastern Massachusetts and discharge to the estuaries of the Seekonk River and Providence River in Rhode Island, across from the city of Providence, eventually flowing into Narragansett Bay (Figure 3.53). Towns in Massachusetts in the Ten Mile River Basin include Plainville, North Attleborough, Attleboro, and Seekonk. Pawtucket, Rhode Island, is also in the basin.

The river system was formed about 12,000 years ago as the continental glaciers receded, yielding slow-flowing streams of glacial purity, with only a slight slope toward the ocean, and creating many natural habitats for stream and wetland animals

Figure 3.54   Arrowheads and stone implements from Woodland Age settlements along the shores of Ten Mile River. (Drawing by William S. Fowler.)

and vegetation. After almost 10 millennia of sparse human habitation by hunting bands, settlements began along the river slightly downstream of what is now the center of Attleboro, according to archaeological evidence. From temporary campsites established by hunters, this favored location became the home of agricultural settlers, at the time of the early Middle Ages in Europe (Figure 3.53). Most of the banks of the Ten Mile River contain evidence of campsite and agricultural activity that occurred before the 17th Century (Figure 3.54).

After colonization of the region by immigrants from Europe and elsewhere, the river was dammed to provide reliable sources of power, and the valley eventually became the home of jewelry and metal-plating industries that used the river for process water and waste discharges. This activity destroyed the natural environment in the river and was the forerunner of severe problems with local floods and environmental contamination.

During the early 20th Century, the industrial discharges created highly acid conditions in the river. In combination with high concentrations of metals and excess organic pollution, this eliminated most forms of aquatic life from the river during the early 20th Century. If environmental quality of the river were to be rated from 1 to 10, the rating for the river during this period should be 1 or zero — the lowest possible value. In contrast, the rating would probably have been 10 for the preceding 12,000 years.

The major population centers of North Attleborough and Attleboro developed sanitary sewer systems and began discharge of partially treated domestic wastewaters soon after World War II. Lack of adequate treatment of these wastes added to the problems in the river, especially because of the low flow in the river system during the summer and fall when critical conditions occur for aquatic life. During the hot summer months, almost all flow in the river is from these wastewater discharges, with the largest portions coming from the municipalities of Attleboro and North Attleborough. During low-flow conditions, the river discharges about 15 cubic feet per second (cfs) or 10 million gallons per day into the Seekonk River estuary. Over 12 cfs of this is from wastewater discharges, with 8.6 cfs from the Attleboro treatment plant (Figure 3.55). With only about 15% of the river being clean water available for diluting the wastewaters, extremely high degrees of wastewater treatment are needed for this system to approach primeval purity.

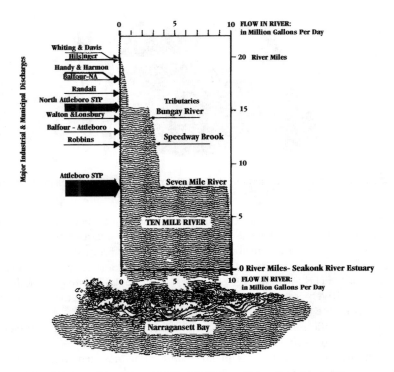

Figure 3.55   Flow pattern in Ten Mile River during drought conditions.

Additional problems have been caused by settling of the organic and metallic materials from the various discharges into the impoundments created by the 15 small dams along the river. Not only have the sediments produced obnoxious conditions in the ponds and eliminated the flood storage capacity of these reservoirs, but they have also locally raised the general river level and water table, causing increased infiltration into sanitary sewers that parallel the river course.

An additional reason to modify or remove many of these small industrial dams is the infiltration and excess flow in sanitary sewers caused by the locally higher water tables. This increased infiltration caused operational difficulties in the municipal treatment plants during rainy periods, and also resulted in loss of useful flow capacity in the sewer systems. Flooding occurs quite frequently in the Attleboro area, probably increased by the dams and siltation.

In the 1960s, a broad program for control of water pollution was initiated in Massachusetts, and high priority was given to the Ten Mile River because of the severe nature of metal and organic contamination. Reduction in gross contamination then occurred relatively rapidly in this river system because of the eventually cooperative attitude among many local industries.

The 1970 field survey of the Ten Mile River was conducted to determine conditions after the gross industrial discharges had been curtailed. It was a joint federal/state survey designed to collect precise data on organic pollution and oxygen depletion in the river from the remaining municipal discharges.

That summer of 1970, we decided to make predawn runs on the river in order to catch the oxygen concentrations at their lowest points of the day. After a full night of respiration, the algae and bacteria have depleted the oxygen generated by sunlight of the day before. Measurement of these daily fluctuations gave us a clear picture of the average oxygen concentrations as well as the daily highs and lows experienced by the fish. We ran the river sampling four times a day for 2 days in mid-July and 2 days at the end of July, 16 runs in all, hitting 26 stations on this tiny river and its tributaries.

The dawn run was the best. The team leader picked me up on the interstate highway about 3 a.m. We had our bottles tagged and log books ready from the day before. We pulled the first samples from the river around 4 a.m. As we approached the river bank, ghostly vapors rose in the dark stillness. Many of the millponds were covered with luminescent fog. It muffled the creak of the oars on our jon-boats as we paddled out to the middle of the mill pond. The hulking, abandoned mill buildings were cloaked in fog, silently watching our careful rituals of sample collection and preservation.

And then came the dawn as we finished the first river run. It was time for breakfast. The crews had their favorite roadside diners on each river. In keeping with the river towns, the diners were grey and drab, but the coffee and hash browns were hot. In a few days, the camaraderie of the the the river crews developed into a tight band. It became *our* river, *our* mission. Later that year when we met in Boston for enforcement proceedings, we relived the summer surveys, relishing the tales of capsized boats, Mac's tyrannical field lab, run-ins with irate residents and dogs, and tales of pickerel and turtles. The talk of river rats.

Over the ensuing two decades, the Ten Mile River showed significant improvement in water quality and in the health of its aquatic biota, especially since the 1970s, when improved treatment of industrial wastewater began. The concentrations of harmful metals were significantly reduced by controlling and treating the industrial wastes with the best available technology, thus creating conditions in which some small aquatic animals were able to survive.

Due to decreased organic pollution since 1980 when the municipalities of North Attleborough and Atttleboro both began operating advanced systems for treatment of domestic wastewaters, oxygen values came back to normal concentrations in many areas, so that fish might return. Although the general appearance and vegetation of the river impoundments remained unattractive due to overfertilization of algae and aquatic weeds, this problem was reduced to more manageable proportions by the installation of nutrient removal systems in these new treatment facilities.

3.4.5.1 Wastewater Discharges in 1984

During 1984, an inspection of the river indicated 21 separate discharges of potentially hazardous wastewaters, and 15 impoundments along the mainstem of the river (Figure 3.56). Sampling of these discharges for chemicals was conducted during the summer in cooperation with the industries and municipalities, and biological assays of the toxicity of the effluents on aquatic organisms were carried out in 1984 and 1985. Copper and lead were found to be the most important contaminations, and many other contaminants were at unacceptable concentrations, despite the existence and generally efficient operation of the treatment facilities constructed in the early 1970s.

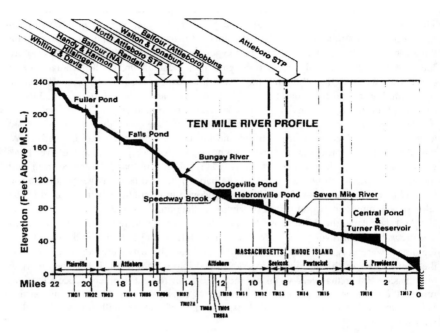

Figure 3.56 Profile of Ten Mile River showing five principal dams and 1984 river sampling stations. River miles begin at zero at the mouth of the river at the right of the drawing. However, the sampling stations run in the opposite direction, with TM01 at River Mile 21 upstream, and TM17 downstream at River Mile 1. Major sewage discharges are indicated at the top of the panel, with the width of the arrows roughly proportional to the quantity of flow in the discharge.

Copper and other metals are dangerous to aquatic life because they can be lethal; and even at sublethal concentrations, they disrupt normal life processes. These metals limit reproduction and stunt the growth of aquatic organisms if continuously present at concentrations in the range of 0.1 to 50 parts per billion (ppb), with silver and lead causing impacts at the lowest concentrations. Rapid death occurs from these metals if aquatic organisms are exposed to higher levels. Copper and silver start causing mortality in a matter of hours if the concentrations exceed 10 ppb.

The total amount of copper reaching the river from the discharges varied from 10 to 35 pounds per day in 1984, about 4 tons per year. Lead and nickel, as well as silver, ammonia, and chlorine, were also present in toxic amounts in some discharges, and large amounts of phosphates, which fertilize algae and other vegetation, were being discharged to the river by the two municipalities.

In many cases, the concentration of contaminants in the industrial discharges were over 10 times the allowable limits for the river, based on chronic toxicity considerations for aquatic organisms or on background levels in clean water (Figure 3.57). For example, 10 of the 16 major discharges contained copper concentrations over 10 times the background limit of 20 ppb. All industrial discharges contained more than 10 times the chronic toxic limit for lead, and background levels

Contaminant & Concentration In Effluent

Discharges	Discharge From Treatment Plant (Thousand Gallons Per Day)	Index of Acidity (pH) (Standard Units)	Suspended Solids (Parts Per Million)	Silver (Parts Per Billion)	Copper (Parts Per Billion)	Nickel (Parts Per Billion)	Lead (Parts Per Billion)	Zinc (Parts Per Billion)	Ammonia (Parts Per Million)	Phosphorus (Parts Per Million)	Chlorine (Parts Per Million)
Whiting & Davis	50	8	12	20	2000	1300	60	240	0.6		0.3
Hilsinger	40	8	4	3	15	0	20	10	0.1		0.2
Handy & Harmon(NA)	130	10	14	1700	2500	7000	60	0	17.0		2.0
Balfour (NA)	25	8	5	13	1600	630	130	140	0.2		0.2
Randall	60	7	8	50	720	460	20	130	0.1		0.2
North Attleboro STP	2000	7	4	0	80	130	20	0	0.1	0.5	1.0
Walton & Lonsbury	30	7	9	0	70	70	0	20	0.1		<0.1
Mt. Vernon Silver	1	10	21	10	1400	0	80	2200	0.0		0.1
Balfour (Attleboro)	43	7	2	30	890	1050	40	340	0.1		0.1
Foster Metal	1	10	5	10	1300	620	90	130	0.1		2.5
Lambert Anodizing	0.2	7	4	0	80	40	100	40	<0.1		<0.1
Leavens	12	7	3	90	12000	1710	100	900	9.0		0.1
Leach & Garner	13	7	3	0	280	50	80	130	<0.1		<0.1
Swank	90	8	2	0	50	340	60	8	0.6		0.2
Robbins	15	10	10	80	3500	240	100	440	4.5		0.1
Attleboro STP	5670	7	1	0	160	15	0	0	0.5	2.6	2.0
Acute Toxic Limit				1	8	1100	25	180			
Chronic Toxic Limit				0.1	6	56	1	47			
Clean Upstream Concentration		7	4	2	20	20	40	100	0	0.2	0

Figure 3.57 Heavy metals and other contaminants in 16 major discharges to Ten Mile River, 1984. Black squares indicate values highly toxic for aquatic life.

were so high that there must have been significant additional sources of lead in this river basin, probably from roadways (including Routes 1, 1A, and several interstate highways) that crossed the basin or paralleled the stream courses (Figure 3.53).

The biological verification of the chemical data was summarized in the "toxic quotient," indicating the relative biological hazard of the discharge after dilution in the river. The most hazardous discharges had toxic quotients around 200, while only a few were significantly below 1, a relatively safe indication (Table 3.11). These quotients were obtained using the sensitive aquatic organism *Daphnia*, exposing groups of them for 48 hours to the discharge being evaluated.

Many of the discharges were toxic even when diluted 100 times with clean water, according to the results from the biological testing. Thus, the use of the treatment technology of the 1970s on these wastewaters was clearly shown to be inadequate for the biological health of the river. More recent federal and state policies require that the protection of the overall biological structure and integrity of the river be the basic goal in developing pollution abatement programs for the nation's waters.

Table 3.11   Summary of Toxicity Results from Two Discharges to the Upper Ten Mile River, 1984

Parameter	Discharge from Whiting and Davis, Inc.	Discharge from Handy and Harmon, Inc.
Flow in gallons per day	50,000	130,000
Dilution ratio available at point of discharge (river flow/plant discharge)	8.2	11.4
Fathead minnow, 50% lethal concentration	25.6%	2.2%
Fathead minnow, no observable effect level	10.0%	0.5%
Daphnia pulex, 50% lethal concentration	0.3%	0.06%
Daphnia pulex, no observable effect level	0.05%	0.05%
Toxicity quotient for fathead minnows	1.08	17.06
Toxicity quotient for *Daphnia pulex*	217	170
Total toxicity units	30	220
Total toxicity units not including chlorine	27	220

Note: Figures given in % indicate the dilution of the wastewater required for the designated lethal effect. For example, if the 50% lethal concentration is 25%, it means that a solution comprising 25% wastewater will give 50% mortality of the organisms placed in it.

3.4.5.2   *Comparative Toxicity Studies*

On one of the industrial discharges of major concern in 1984, detailed evaluations were made of the relation of biological assays of the discharge toxicity, and conditions observed in the macro-invertebrate communities in the stream below the point of the industrial discharge.

The discharge tested was from the electrofinishing plant of Whiting and Davis, Inc. in Plainville, Massachusetts, slightly upstream of Wetherell Pond. At this location, the Ten Mile River is a small stream with a drainage area of 3.3 square miles.

Sampling stations in the river were located 75 feet upstream of the plant discharge, and 150 feet downstream, both stations being upstream of Route 106. Whiting and Davis, Inc. manufactured costume jewelry at their Plainville facility. Wastewaters from the plating processes were segregated into cyanide and non-cyanide streams. The cyanide stream was treated with a two-stage chlorine destruction system, and then mixed with the non-cyanide wastes. The combined waste was then given physicochemical treatment, including storage in a lagoon, which then discharged to the river.

The chemical analyses of the discharge from the Whiting and Davis facility indicated that copper and chlorine might be at high enough concentrations to cause toxicity to aquatic organisms (Table 3.12).

In addition to the chemical analyses, biological measurements of toxicity were made, exposing fathead minnows and the water flea *Daphnia pulex* to the wastewaters for 48 hours, with small amounts of the wastewater being added to a stock solution of clean water. By this method, it was possible to determine the percentage of wastewater that would produce 50% mortality among the test animals, and also

Table 3.12 Chemical Analyses of Discharges from Whiting and Davis Industries

Parameter	Units	Measured September 1984	NPDES permit limit	Measured December 1982	Measured September 1982
Total suspended solids	Parts per million (ppm)	12	20		
pH	Standard units	7.8	6.0–9.5		
Ammonia	ppm as N	0.55			
Unionized ammonia	ppm as N	0.0132			
Copper	ppm	2.0	1.5	0.42	0.06
Zinc	ppm	0.24		0.11	0.25
Total chromium	ppm	0.02	1.5	0.02	0.02
Lead	ppm	0.06			
Nickel	ppm	1.3	1.8	0.25	0.53
Silver	ppm	0.02	0.15	0.05	0.13
Cyanide	ppm	0.02	0.25		0.01
Cyanide-A	ppm	<0.01	0.1		0.01
Temperature	°C	21.5	23.9		
Chlorine	ppm	0.3	0.2		
Aluminum	ppm			6.9	1.5

the percentage of wastewater that would have "no observable or adverse effect," an EPA designation for toxicology studies. These analyses were conducted for both the Whiting and Davis discharge and also a discharge from another electrofinishing plant further downstream, Handy and Harmon of North Attleboro. The Handy and Harmon discharge entered the Ten Mile River directly upstream of Falls Pond, at River Mile 18.1 (Figure 3.56).

The dilution of plant discharge provided by the Ten Mile River was about 10:1 for the flows of 50,000 and 130,000 gallons per day during 1984. This dilution with relatively clean and nontoxic river water should have provided some protection for the aquatic organisms. However, the laboratory testing showed that more than 10% of the wastewater in a stock volume of clean water would cause adverse effects. If the amount of wastewater from Whiting and Davis added to the clean water reached 26% of the stock volume, half of the fathead minnows would die (Table 3.11). If only 10% of the wastewater was introduced to the stock volume, none of these minnows would die.

Similar toxic assessment of the discharge from Handy and Harmon indicated that no observable effect would occur until the amount of wastewater became greater than 0.05% of the stock volume, and that half of the fathead minnows would die if the wastewater volume were 2.2% (Table 3.11). As sensitive as the fathead minnows were, the *Daphnia pulex* showed even greater susceptibility to copper and other toxic materials in the wastewaters from the two facilities. The 50% lethal concentration for *Daphnia pulex* from copper in the discharge of Whiting and Davis was 0.3%, and for Handy and Harmon it was 0.06% (Table 3.11).

This numerical summary of toxic effects in the two industrial discharges showed that the discharge from Handy and Harmon, Inc. was much more toxic, with 220 toxic units, compared to 27 to 30 toxic units in the discharge from Whiting and Davis, Inc.

To corroborate the chemical analyses and the laboratory assays of biological toxicity for the Ten Mile River downstream of the Whiting and Davis discharge, the communities of aquatic organisms were surveyed in the field during the summer of 1984, at about the time that the other measurements were made.

Frequency distributions of individuals were quite different at the upstream and downstream stations, except for the chironomid fauna. Tubificid worms were a major component of the upstream community, but were completely absent downstream. The aquatic sow bug *Asellus communis* was widely and evenly distributed upstream, but much less abundant downstream.

The distribution of aquatic mollusks was affected strongly by the discharge. Species of the common *Gyraulus* snails were found widely distributed upstream of the discharge. Altogether, there were four taxa of mollusks upstream, but none downstream.

There were both fewer numbers and fewer taxa downstream, in the non-chirono-mid organisms. The consistency in the differences between community indices of upstream and downstream collections, the differences between these collections in the better represented taxonomic groups, and the drop in total number of individuals downstream indicated that the downstream community was stressed.

3.4.5.3   *Water Quality of the River in 1984*

The water quality results from the 1984 surveys were somewhat encouraging. The acid level in the river was normal, with a mean pH slightly above 7, which is highly suitable for aquatic life. This contrasted sharply with the low pH from 2 to 5 found in a survey during the Great Depression (Figure 3.58). The low ph in 1937 was due to acid wastes from the metal and jewelry industries. Acidity had been brought to near normal by 1973 when most of these industrial wastes were given the best treatment available at that time, as detailed by an earlier survey of the river.

Suspended solids concentrations had also been high back in 1937, between 10 and 20 parts per million (ppm). This had become worse by the time of the 1973 survey because additional solids came from the municipal sewage of North Attle-borough and Attleboro, reaching a mean of 14 ppm, and exceeding 26 ppm at River Mile 8 below the discharge from the city of Attleboro (Figure 3.59). However, in 1980, new treatment plants began operating in both municipalities, reducing the suspended solids in their discharges to below 5 ppm. The results were evident in the low suspended solids found in the river in 1984, nearly the same as the 5 ppm found in clean water upstream of all discharges (Figure 3.59). In 1990, the concentrations of suspended solids were even lower, except at the outlet of Hebronville Pond at River Mile 10, due to excessive algae production in the pond.

Metals, even after dilution in the river, were at very harmful concentrations in 1984, especially copper and lead (Figure 3.60). Under normal conditions, adverse impacts on aquatic biota would occur at concentrations of copper above 6 parts per billion (ppb), and this value was exceeded at every sampling station on the river, reaching maxima over 100 ppb. Other metals found in harmful concentrations included lead at 40 to 100 ppb and nickel at 10 to 80 ppb. The mean concentration

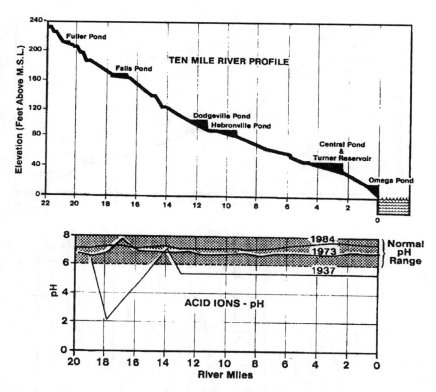

Figure 3.58   Historical changes in acidity as measured by pH units in Ten Mile River, 1937–1984.

of lead in the river was over 60 ppb, significantly higher than the 20 ppm recorded in 1981. This increase was not entirely due to industrial discharge, as the lead concentrations in the clean water stations, upstream of any discharges, were also high in 1984. They averaged 40 ppb in 1984, compared to virtually zero in 1981 (Figure 3.60).

3.4.5.4 Metals in Sediments

The surveys in the impoundments showed high concentrations of metals in the sediments, and in some areas, no bottom organisms (Table 3.13). Copper and nickel content exceeded 2000 ppm in the sediments. When biologically tested for toxicity, the sediments showed severe adverse effects on biota, especially in the impoundments downstream of Attleboro center. In Central Pond, the sediments showed toxic impacts on microorganisms until the sediment was diluted to 8% of its natural strength. This is reported as the sediment toxicity index, the dilution of sediment required before 50% of certain microorganisms can survive a 30-minute exposure (Figure 3.61). A large index such as 100% indicates the sediment is only mildly toxic, whereas a small index of 1% or 5% indicates high toxicity.

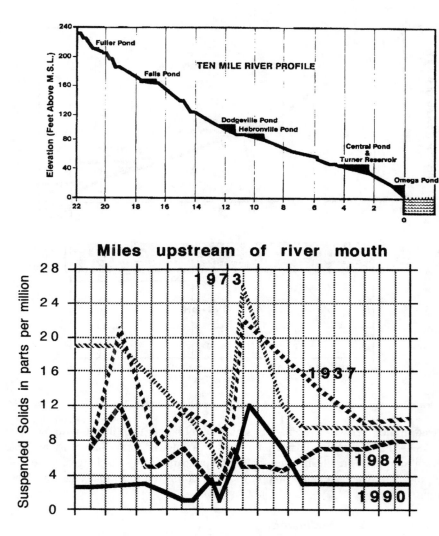

Figure 3.59   History of suspended solids in Ten Mile River, 1937–1990.

The high concentration of lead in sediments of the upstream ponds and then the doubling of this concentration in sediments near the river mouth indicated contamination derived from transportation sources, reaching serious levels. There were no industrial discharges above Fuller Pond where the 100 ppb concentration of lead was found. Copper and nickel may also be partly derived from bearing wear and other transportation sources, but the observed concentrations in reservoir sediments is closely related to industrial discharges, and the amount in sediments of the upstream Fuller Pond was only a few percent of the maximum amounts found in Dodgeville and Central Ponds, downstream (Figure 3.62).

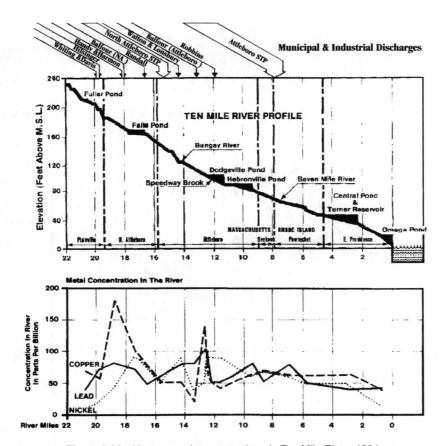

Figure 3.60 Heavy metal concentrations in Ten Mile River, 1984.

Table 3.13 Heavy Metals in Sediments of Reservoirs on Ten Mile River, 1984, in ppm

Pond	Cadmium	Copper	Nickel	Lead
Fuller	2	108	27	107
Falls	12	792	191	176
Dodgeville	219	3230	1600	192
Hebronville	120	773	950	180
Central	166	2680	2160	201

3.4.5.5 Fish

Although warm-water fish were present in small tributaries of the river system, such as chain pickerel and red-fin pickerel (Figure 3.63), the number of species and the numbers of fish were far less than would be expected, based on the size and geometry of the river system.

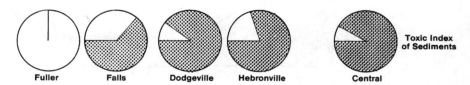

Figure 3.61   Metal concentrations and toxic indices for sediments in major impoundments on Ten Mile River, 1984.

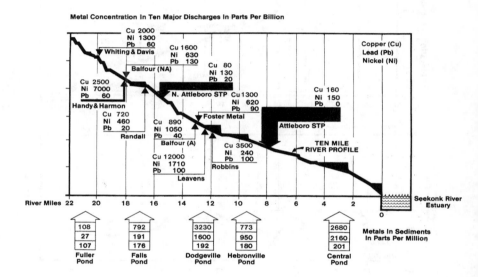

Figure 3.62   Heavy metals in major discharges into Ten Mile River, 1984. Cu, copper; Ni, nickel; and Pb, lead. Concentrations of metals are for total metals, in parts per million (ppm).

Central Pond and Turner Reservoir in Seekonk were the only ponds that had a fairly normal number of fish species, still less than a dozen (Figure 3.64). Even these fish populations were limited in numbers and their flesh contained noticeable amounts of heavy metals, especially lead. Further downstream, oxygen concentrations were observed below the 5 ppm standard needed to avoid fish deaths in the summer. Concentrations of 2, 3, and 4 ppm were found at times, throughout the reaches of the river below Attleboro (Figure 3.64). It was disturbing that the relatively clean upstream impoundments such as Fuller and Falls Ponds also contained limited fish communities. They may have been limited by background levels of copper and lead preventing normal reproduction and survival. However, the richness index was at a maximum of 24 groups of food organisms in this part of the river, indicating a generally favorable ecology.

Measurements of growth of largemouth bass, bluegills, white perch, and pumpkin-seed indicated generally poor conditions, except at the upstream station. A particularly sensitive measure of toxicity is the survival and deformity of larval minnows. This was

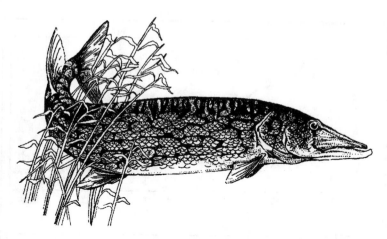

Figure 3.63 Chain pickerel from the Ten Mile River. The cleanest portions of the Ten Mile River supported populations of this predatory fish, indicative of a broad-based food supply.

evident in data for August 1984 that included larval survival of fathead minnows, as well as measurement of numerous suspected toxicants in the Ten Mile River. In clean water at River Station TM01, 87% of fathead minnow larvae survived and appeared normal after 8 days from hatching of eggs. However, at River Station TM03 downstream of the industrial discharge from Whiting and Davis in Plainville, only 30% survived and appeared normal. From 100 eggs exposed to water from downstream, 31 died and 39 were deformed, leaving only 30 normal and alive after 8 days. The principal toxic chemicals in the Whiting and Davis discharge were silver, copper, nickel, and chlorine. In the river water, the silver, copper, and nickel concentrations were 10, 25, and 15 ppb, respectively. Also, the concentrations of lead, cadmium, and ammonia were 250, 20, and 70 ppb, respectively, at this station in August.

For a similar evaluation downstream of the discharges from the jewelry and finishing plants of Swank Co. and Robbins Co. on Speedway Brook (a tributary of the Ten Mile River in Attleboro), the number of dead minnows in water from River Station TM10 was 56 and the number deformed was 15 at the end of 8 days, compared to 8 dead and 5 deformed in the clean water. The principal toxic chemicals in these two discharges were silver, copper, ammonia, and chlorine. In the river at this station, the concentrations of silver, copper, nickel, and ammonia were 5, 25, 170, and 140 ppb, respectively. The concentrations of lead and cadmium were 160 and 25 ppb, respectively, in the river at this point.

There seemed to be a direct correlation between two of the metals and the impact on the minnow larvae, whereas the other contaminants seemed unrelated. Silver appeared to be directly related to deformities in the larvae, while cadmium seemed to be directly related to death. The potential impact for these relationships on the normal fish populations is very significant. The sensitivity of the eggs and larvae of fish to these metals indicates that the larvae should also be useful for monitoring biological impact of metals.

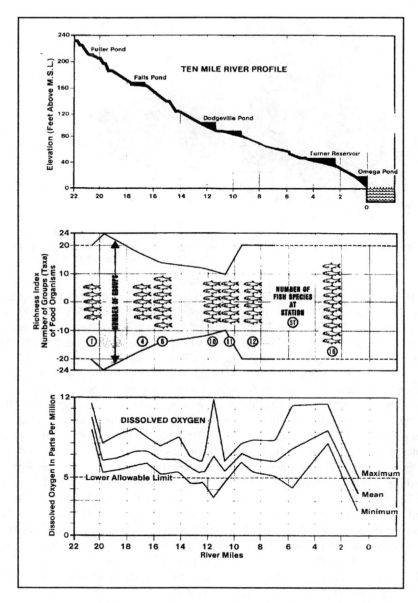

Figure 3.64   Number of fish species, related richness index for food organisms, and dissolved oxygen concentrations in Ten Mile River, 1984.

Metals were found in the flesh of bluegills and white perch, indicating that their food source — which is usually algae — may have accumulated heavy metals from the river waters. Although the upstream clean water station at River Mile 21 had low concentrations of metals (Figure 3.65), there was a significant accumulation of

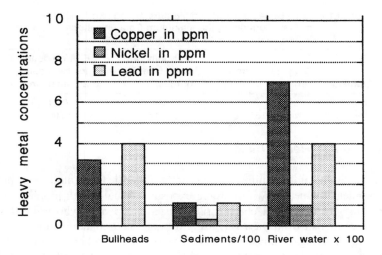

Figure 3.65   Heavy metals at clean water station above Fuller Pond at River Mile 21 on Ten Mile River, 1984. Compare this figure with Figure 3.66, which shows heavy metals in contaminated pond downstream.

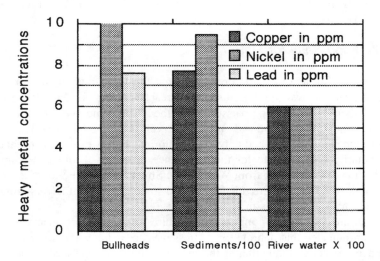

Figure 3.66   Heavy metals downstream of several discharges from metal finishing plants, in Hebronville Pond at River Mile 9 on Ten Mile River, 1984. Compare this figure with Figure 3.65, which shows concentrations of metals from upstream clean station.

metals in bullheads from Hebronville Pond in Attleboro at River Mile 9 (Figure 3.66). Copper, nickel, and lead were found in the flesh of bullheads at 3, 13, and 8 ppm, respectively, in Hebronville Pond. Bullheads browse the bottom muds where the metals were concentrated.

Table 3.14   Daily Doses of Lead and Mercury, Using Two Meals per Week
of Fish Containing the Maximum Concentrations Observed
in Ten Mile River Specimens

Element	Daily intake from eating fish, in mg	Total daily oral intake from all sources, in mg	EPA guideline for total daily oral intake, in mg
Lead	1.2	1.6	0.31
Mercury	0.018	0.018	0.020

3.4.5.6   Human Health Risk

A risk assessment for families living near the Ten Mile River indicated a potential health problem for people with chronic exposure. Acute health problems did not appear likely. The potential chronic risk existed for people who might take one weekly meal of fish from the river, using average lead contamination values. For urban children living in houses with high amounts of lead-containing paint, and drinking water with lead from the plumbing system, the risk would be serious.

If the maximum total oral intake of lead from fish is calculated using two meals of fish per week and the maximum lead concentrations observed in the fish, the daily intake of lead would be over 5 times the amount recommended under EPA guidelines. This EPA-recommended amount in 1983 was 0.31 milligrams (mg) (Table 3.14).

3.4.5.7   Food Base for Fish

The variety of small aquatic organisms that form the food supply for fish and other larger animals was quite limited in the reaches of the Ten Mile River passing through urban areas, in comparison with the clean water communities found upstream and in Central Pond and Turner Reservoir (Figure 3.64). The river in North Attleborough and Attleboro had only half the normal variety of organisms, and these were primarily groups noted for their tolerance to metals. Biological community structures in the entire river from West Bacon Street in Plainville to its terminus in the Seekonk River in East Providence showed some evidence of adverse impact, in comparison to healthy rivers in New England.

Careful analysis of the community structures of aquatic organisms in the Ten Mile River indicated that definite toxic stress was occurring where the river crossed West Bacon Street in Plainville, at the inlet to Falls Pond in North Attleborough, and in Speedway Brook in Attleboro. Below and even above the two municipal treatment plants, the aquatic communities were "filter-feeding" assemblages, typical of streams receiving high loads of particulate organic matter. At the Olive Street crossing south of Attleboro Center, the structure of the biotic community indicated a combination of toxic stresses and an overly rich organic environment.

The general surroundings and real estate in the upstream reaches of the river near Plainville were markedly more appealing than those of the impoundments downstream of the wastewater discharges (near Seekonk and into Rhode Island) giving graphic and persuasive proof of the damage to property values being done by the wastewaters. The impact of this pollution on esthetic and economic values could also be seen in the type of residential development along the river, with spacious single-family homes upstream, compared to crowded developments along the banks in the downstream portions of Attleboro. Rehabilitation of these old riverside mills for housing or other uses will not really be successful until the river itself is rehabilitated.

3.4.5.8 Nutrients

A major impediment to the final restoration of the river was the large residual of fertilizing phosphates in the discharge of the municipal sewage treatment plants. Simple application of 1970 technology for phosphate removal resulted in discharges containing 1 to 3 ppm phosphates, compared to upstream values of 0.2 ppm, and truly desirable values below 0.05 ppm. Thus, the river in 1984 contained excessive phosphates, lower than those of 1973 before the new municipal treatment plants began operation, but still high enough to stimulate excessive growth of aquatic vegetation and algae (Figure 3.67).

A primary effect of the excessive vegetation resulting from overfertilization by phosphates was seen in the wide fluctuations in oxygen in the impoundments, often dropping to 2 or 3 ppm just before dawn, which is dangerous to fish. Another important effect of rooted vegetation is that it decreases local velocities of flow, especially near the edges of a pond, thus promoting stagnation and increased sedimentation in those areas.

The effect of the phosphate removal systems installed around 1980 showed clearly in the drop in chlorophyll-a concentrations in the river from 1973 to 1984 (Table 3.15) The amount of chlorophyll-a is a good quantitative measure of the mass of algae present, even better than counts of actual algae cells, which are highly clumped and produce very erratic numbers. The drop in chlorophyll between 1973 and 1984 from means of 36 ppb down to 10 ppb (72% reduction) occurred in parallel with the drop in phosphates in the river from means of 0.82 ppm down to 0.35 ppm (57% reduction), due to the new municipal treatment plants.

It is possible but expensive to further lower the concentrations of phosphates in the river to the required values. It would require not only improved removal in the municipal plants, but also control of seepage from septic systems and lawn fertilizers in residential areas. These latter items contributed to the high background of phosphates at the clean water stations.

3.4.5.9 Analysis

Although noticeable progress had been made in restoring the Ten Mile River during previous decades, the river was still receiving 2 to 8 tons of copper annually from wastewater discharges in 1984. Fish had returned to the downstream impoundments on the river but their flesh contained undesirable amounts of lead, copper, and other metals.

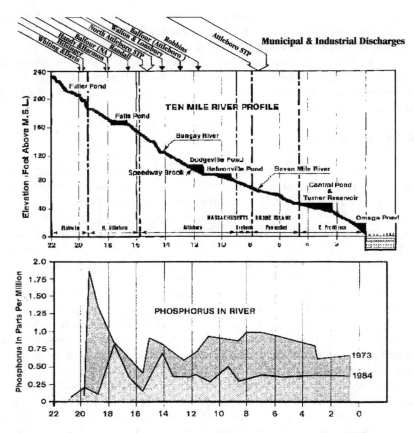

Figure 3.67   Total phosphate nutrients from Ten Mile River, 1973–1984.

Table 3.15   Concentrations of Chlorophyll-a and Numbers of Algae Cells per milliliter in Ponds of Ten Mile River, Summers of 1973 and 1984

Pond	Chlorophyll-a 1973, in ppb	Chlorophyll-a 1984, in ppb	Number of algae cells per milliliter, 1984
Fuller	5	1	31
Wetherells	40	2	592
Falls	95	15	14,525
Farmers	40	18	9,241
Mechanics	30	2	295
Dodgeville	15	25	571
Hebronville	20	1	168
Ten Mile Reserve	25	2	255
Turner		20	5,324
Central Pond	50	21	5,406
Mean	36	10	

Lead concentrations in tadpoles living near highways may contribute to the elevated lead concentrations reported in fish and other predators on the tadpoles. Some birds and mammals have shown physiological and reproductive impairment due to dietary sources of lead coming from tadpoles, minnows, and other aquatic organisms.

Field observations on ponds and drainage systems in the state of Maryland showed that lead concentrations in tadpoles and in sediments were closely related to traffic volumes on nearby highways. This would explain the high lead concentrations in the upper reaches of the Ten Mile River that pass near so many interstate highways and main roads.

Accumulation of metals in the organic sediments of the several impoundments presented a hazard not only for aquatic life in the river, but also for the estuarine and marine life downstream in Narragansett Bay, a highly attractive recreational and natural resource area. Shellfish, which are common in Narragansett Bay, may bioaccumulate copper by factors of 20,000 to 30,000. Thus, the metal problem in sediments of the Ten Mile River impoundments might also have widespread impact across state lines after heavy rainfalls and scouring floods. Hurricanes have hit the Narragansett Bay drainage basin often in the 20th Century and probably have important influences on the distribution of heavy metals in the sediments of Narragansett Bay.

The conclusions drawn from chemical analyses of the river were confirmed by biological assays of the wastewater discharges and the river, and by examination of the structure of biological communities under toxic stress from excessive amounts of copper and other metals. Aquatic organisms, because of their small size and total immersion in water, are even more sensitive to copper and other metals than are people, when the standards are expressed in terms of mass of metal per volume of water. Thus, the standards for metals in the river must be more stringent than drinking water standards for human consumption. Human consumption of metals in drinking water is limited by the amount of water a person can drink, a relatively small amount in comparison to human body mass.

Analysis of the data also pointed to a problem in the downstream impoundments with excessive nutrients and vegetation. The excessive algae and vegetation in impoundments in the downstream portions of the river valley caused continuing problems with maintenance of proper oxygen levels in the river, a prime necessity for fish survival. Refinements in the operation of municipal treatment plants in North Attleborough and Attleboro could bring phosphate concentrations down to more acceptable values and raise mean oxygen levels, thus stabilizing oxygen conditions for fish and probably improving metal removal at the same time.

Although the fish populations were abnormally limited and had elevated levels of metal in their tissues, it is likely that they can be brought to a healthy condition by further control of metal discharges and subsequent restoration of the river biota upon which they depend for food. However, the increasing concentration of lead in the river since 1981, and the likely high levels of lead already in the blood of people in urban areas of the lower basin, indicate a present potential for public health problems in addition to the dangers to aquatic life.

3.4.5.10 Perspective

For the woodland hunters who gathered at campsites several hundred years ago along the Ten Mile River, a reverent stewardship of the land and water was part of their way of life. The original North Americans who fished these rivers would not pollute or damage them, the rivers being their source of food, drink, and cleanliness.

But the concept of stewardship of the river escaped the immigrant and industrialized society that displaced the original North Americans. Americans dammed the river in the 19th Century to extract energy for mills and metal shops. After 100 centuries of crystal clarity and biological harmony, the river was converted into an "acid bath" in a few, short years.

> Mills had to be built. And so The River began to die. Children...abandoned one stinking pool after another. Ponds where trout and bass once lurked...yielded bullheads, then nothing. Trash accumulated... It became no more than an open dumping ground."
>
> Rumford River Laboratories, 1985

As the American economy recovered from World War II and people had time again to sit on the banks of the rivers and to row boats on the mill ponds throughout America, they were appalled at the desecration. Their awareness gave birth to a new concept among those responsible for our public health and our environment: that engineers had to design wastewater treatment plants to fit the needs of the river, not just to satisfy short-term economic considerations. The river was more than a stream for dilution of wastes; it was a complex and balanced community of water, plants, and animals that required protection. Action finally began after Earth Day 1970, based on this concept.

By 1973, the acid level of the Ten Mile River was again tolerable to fish, and in 1984 it had returned to a normal level, suitable for living things. This improvement took a full generation, but it is hardly finished. Residues of waste remain; some are still being poured into the river. So the movement must continue.

To quote again,

> ...it does not have to stay dead. Now...we see the incalculable worth of our vanishing natural resources. We see the vital necessity for clean water... It is more than being able to swim and fish and picnic in nearby green acres... It is a value that exists in the heart...whose name is civic pride. Only this...can make The River live again.
>
> Rumford River Laboratories, 1985

3.4.5.11 Enforcement Experiences and the Death of the Green Movement

There were two distinct phases in recent efforts to clean up the Ten Mile River, beginning with the "first flush of excitement" preceding Earth Day 1970. This first phase was chaos. The federal and state governments had reformed their environmental agencies, and Massachusetts had a sparkling new Division of Water Pollution Control (DWPC) ready to clean up the rivers of the Commonwealth.

First Flush Round — As in all new organizations, these new environmental agencies had clear missions, but little experience. As the first round of river surveys began, and attempts were started to reduce the major polluters, the Ten Mile River was picked because of its unusually severe metal contamination. About 1965, the industries discharging these metals and other highly toxic wastes were asked to install industrial treatment facilities. One at a time, they were brought in for enforcement conferences, and told that their discharges had to meet the new water quality standards.

The industries resisted furiously. They threatened to leave Massachusetts; they threatened the DWPC with legal action because it was unfairly targeting them while other major polluters were getting away with murder; they insisted they could not operate if they had to treat their wastes; they claimed they would be driven into insolvency; they accused us of trying to ruin the Massachusetts economy; and they promised to go to their legislators and to the governor.

The threats and arguments became familiar over the years, but there were only so many they could invent. However, in this first round, every step required enormous effort. After several years, most of them had installed at least the simplest form of treatment, but only because of days and weeks of diligent pressure by technical and legal staff of the new DWPC. Everyone was new at the job, the regulations were new and not always clear, and there were cases where DWPC demands were not technically sound because of its inexperience. It was a long decade.

By the end of the 1970s, the DWPC began to see some real improvements in the river. Industrial discharges were getting treatment but they were also switching processes; thus, there were other problems with ammonia and chlorine that had not existed before. The municipalities had not improved their treatment facilities, so organic pollution was still severe. Thus, it was decided that a second phase of the Ten Mile River cleanup was needed. But this time, the DWPC knew what it was doing.

Brass Knuckles Round — In the second phase — called the "brass knuckles round" — the state and federal agencies had an excellent working relationship after a decade of field surveys and improved laboratory facilities. The DWPC had adequate field personnel through special summer employment programs; its legal people had dealt with all of these towns and industries before, and knew all the games.

In 1984, the latest field data had been analyzed and new biological toxicity measurements had been added to the previous chemical sampling programs. The DWPC sampled for several days instead of two or three. The industrial facilities were visited and inspected. Permit requirements were analyzed.

Finally, the big day came for enforcement. But this time, there was one enforcement conference, to which all the industries were invited. The environmental agencies had the U.S. congressman and state legislators on hand, as well as all the top federal and state environmental scientists. The field data were given out, explained in detail, and violations clearly noted.

The highly organized "brass knuckles round" paid off. Although a few spokes-men for the industries made feeble attempts at self-justification after DWPC pre-sentations, they clearly had been put on the defensive. It was found that the publicity about the enforcement actions had gotten to the local schools and school children, so these company presidents and chief engineers had heard it from their kids that morning at breakfast — that it was time to stop polluting the river. When they also got it with brass knuckles at the conference, it was too much for them.

Implementation of this second phase went quickly. Within 2 years, all of the industries had either installed pretreatment plants that were then connected to the municipal sewerage systems, or they had redesigned their own plants to recycle all of the wastewaters. The total number of industrial discharges into the river went from 27 in 1984 to 2 in 1990, and the remaining discharges were mostly cooling water with few pollutants.

The effectiveness of governmental efforts to restore the rivers was clearly related to the experience and resources of the enforcement agencies. Such experience took about a decade to accumulate, and for interstate rivers it included the chance for several state agencies to work in consort with the federal EPA, both in making the field measurements needed for enforcement or court action and in coordinating that action. Sufficient numbers of qualified personnel with a decade or so of experience working together were also needed. During the initial growth phase of the Green Movement, the water agencies needed about a decade to reach this critical mass of enforcement experience in order to deal effectively with the determined resistance of industrial and municipal polluters.

The End of the Green Movement — Unfortunately, the Green Movement did not last. We won the battle but lost the war. Precisely because of the effectiveness of the "brass knuckles round," the industrial resistance shrewdly moved its focus to the state legislatures and the federal congress about 1990. Thus, instead of fighting enforcement actions against them due to their contamination of a particular river, the industries banded together and cut the budget and staffs of the enforcement agencies. The impact on environmental agencies was rapid; by 1995 the movement was over. This cyclical ebb and flow of efforts to control water pollution can be numerically summarized in terms of the several studies of the Ten Mile River in recent decades.

At the beginning of the Green Movement, about eight people were involved in the 1973 field survey of the Ten Mile River. They spent about 1 month in the preparation, execution, and reporting of the study. As the effort expanded and improved, over 40 people were involved in the next survey of 1984, including an outside consultant to handle reporting and public information. This combined state and federal effort lasted over 6 months. However, as resistance to the Green Move-ment began to exert itself, the field organizations were pared down by frequent budget cuts and reorganization.

The most recent survey of the Ten Mile River was limited to a modest effort, involving only 10 people who worked for less than 1 month. It is quite possible that this will be the last field survey for at least another generation — until the Ten Mile River again becomes an open sewer and people revolt against the desecration, demanding new studies and improved treatment. The most recent cycle of sanitary awakening and decline took about 30 to 40 years.

At the end of the 20th Century, most of the industries on the Ten Mile River were recycling or conserving their wastes, and were connected into municipal treatment plants instead of discharging directly to the river. Despite this apparent progress, serious problems remained with the contaminated residual sediments in reservoirs. For the most part, the industries no longer used river water, and the dams were abandoned and ignored. The reservoirs were gradually filling in, trapping the noxious sediments under layers of organic silt and roadway debris. A large part of the watershed was paved over with roads, parking lots, and large buildings.

The difficult question of what to do about the toxic sediments had not been faced by any of the environmental programs. If left in place, they posed the risk of downstream contamination after hurricanes. If removed, they would be expensive to properly treat, given the large volume of sediments and varied types of toxic materials.

The few green spaces remaining by the turn of the century were near densely populated residential areas and subjected to heavy use. Nonetheless, opportunities for enjoyment of fishing and of observing wildlife were meager because of the impact of the dams and their toxic sediments.

3.4.6 Blackstone River

The Blackstone River begins where Kettle, Tatnuck, and Mill Brooks converge within the city of Worcester, Massachusetts (Figure 3.68). The total drainage area of the Blackstone River in Massachusetts is 373 square miles. It then flows through the northern part of Rhode Island, discharging near Pawtucket, Rhode Island, into the Providence River, a tidal extension of Narragansett Bay. The drainage basin within Rhode Island is 165 square miles, the total basin being about 540 square miles in area.

The Blackstone River is an important natural and recreational resource, and in 1986 the federal government established the Blackstone Valley National Heritage Corridor along the river. It is the second largest river bringing fresh water into Narragansett Bay.

Fourteen significant dams impede the flow of the mainstem of the Blackstone River and several of these dams are currently used for hydropower generation (Table 3.16).

The river has a long history of pollution. Textile pollution began with the establishment of Slater's Mill in Pawtucket, the first site of cotton processing in New England in 1850 (Figure 3.69). Later, steel processing, wire extruding, and metal finishing industrial enterprises used the river for power and waste disposal. Since Earth Day 1970, many of the industries and municipalities have improved the treatment of their wastewaters, but the river continues to suffer from several pollutants.

In 1973, there were five municipal sewage discharges into the Blackstone River and tributaries in Massachusetts, from Worcester, Millbury, Northbridge, Upton, and Hopedale. In addition, sewage from the towns of Burrillvile, Smithfield, and Pawtucket flowed into the river in the Rhode Island portion. In 1973, the suspended solids in the river reached a peak of 75 ppm below Worcester (Figure 3.70).

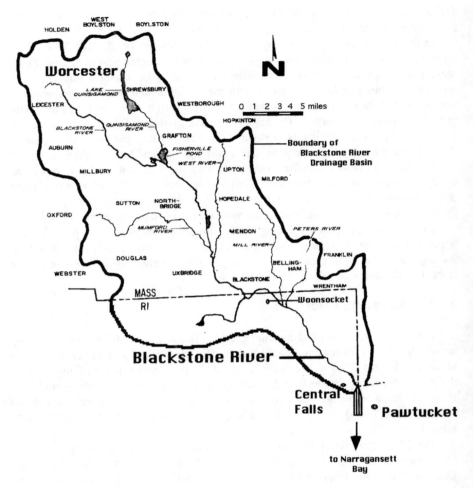

Figure 3.68 Blackstone River Basin in Massachusetts and Rhode Island.

By 1991, the sewage treatment systems had been slightly reconfigured with higher degrees of treatment, resulting in much lower concentrations of suspended solids. However, population growth had resulted in a continuing discharge to Narragansett Bay from all the communities and industries in the Blackstone River basin of several million pounds per year of suspended solids.

There were 18 major wastewater discharges to the river system in 1993, all of them monitored under the National Pollution Discharge Elimination System (Table 3.17). The industrial flows were fairly small in comparison to the municipal systems. Many of the industries in the valley were also tied into municipal sewers and combined their wastes with the domestic sewage before treatment. The largest wastewater discharges in the river system were from the Upper Blackstone Water Pollution Abatement District at River Mile 44.4 in Millbury, and the Woonsocket municipal sewage plant at River Mile 12.5. Together, the flow from these municipal

Table 3.16   Dams on the Blackstone River

Name of dam	Town	River mile
Singing Dam	Sutton	39.8
Fisherville Dam	Grafton	36.5
Farnumsville Dam	Grafton	35.6
Riverdale Dam	Northbridge	31.9
Rice City Pond Dam	Uxbridge	27.8
Uxbridge Dam	Uxbridge on Mumford River	25.5; 0.6
Tupperware Dam	Blackstone	17.8
Thundermist Dam	Woonsocket	14.3
Manville Dam	Cumberland	9.9
Albion Dam	Cumberland	8.2
Ashton Dam	Cumberland	6.8
Central Falls Dam	Cumberland	2.0
Pawtucket Dam	Pawtucket	0.8
Slater's Mill Dam	Pawtucket	0.2

sewage discharges was 67.5 cubic feet per second (cfs), compared to the expected low weekly mean flow at Woonsocket of 101 cfs during a 10-year drought. Thus, in a dry summer, over two thirds of the river flow would be sewage.

3.4.6.1   Phosphates

The concentration of total phosphates in the Blackstone River in the summer of 1973 reached 2 parts per million (ppm) at River Mile 46, below the discharges from the Worcester area (Figure 3.71). This excessive concentration of phosphates continued throughout the river course. By 1991, there was little reduction in nutrients in the river because nutrient removal processes had not been included in the new sewage treatment plant constructed in Worcester in 1978. Ortho-phosphates, the most soluble fraction of the phosphates, were found at almost 1 ppm at River Mile 42 in 1991 (Figure 3.71). Such overfertilization of the river caused excessive algae and vegetation growth down into Rhode Island and Narragansett Bay.

3.4.6.2   Dissolved Oxygen

In the summer of 1973, the dissolved oxygen frequently reached a concentration of zero during the dawn surveys on the Blackstone River, especially below the Worcester area where large organic loadings added to the nighttime deficits caused by overfertilization of the river (Figure 3.72). Improved treatment of the sewage in the Worcester area after 1978 caused improvements in the amount of dissolved oxygen found in the 1991 survey. The mean concentration in 1991 was generally 1 to 3 ppm higher than the standard of 5 ppm prescribed to protect fish.

A permanent monitoring station was established in the Blackstone River at the Rhode Island state line in 1969. Among other parameters, this device continuously measured the dissolved oxygen in the river, making it possible to determine the percentage of time the oxygen was below the desirable concentration of 5 ppm for protection of fish (Figure 3.73). Analysis of the data from the 12 years of monitoring indicated

Figure 3.69 Dam on Blackstone River at Slater's Mill in Pawtucket, Rhode Island.

that the concentration of dissolved oxygen fell below 5 ppm 81% of the time in the summers of 1969 to 1977, before the new treatment plant was constructed for the Worcester area at River Mile 44.4. After 1978 when the new treatment plant began operation, the oxygen fell below the minimum limit only 20 to 30% of the time.

This was a significant improvement in a statistical sense, but a disappointment in an environmental sense. With all of the effort and expenditures of the 1970s to clean up the river, the dissolved oxygen content was still not safe for fish populations.

3.4.6.3 Toxicity from Metals

Extensive studies were conducted on the Blackstone River in 1993 to determine the toxicity of the river and the sediments in impoundments. The biological impact

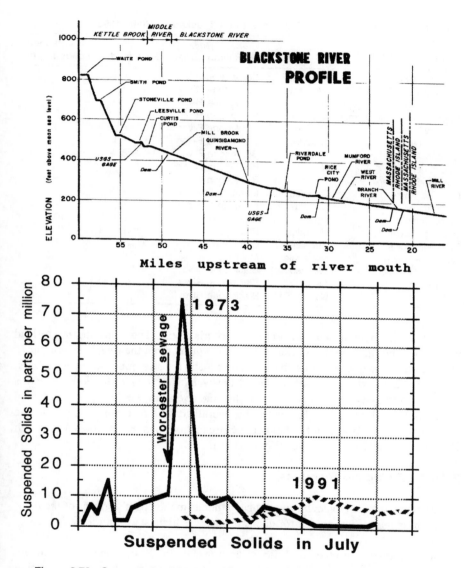

Figure 3.70 Suspended solids concentrations in the Blackstone River, 1973–1991.

of the chemical contaminants was somewhat erratic and generally less than expected, based on EPA criteria for toxicity.

Although several heavy metals were dissolved in the Blackstone River, only cadmium, copper, and lead were found at concentrations toxic to aquatic life. Cadmium was found at concentrations of 4 ppb at River Mile 43, much higher than the EPA criteria for chronic toxicity of 0.7 ppb (Figure 3.74). It remained above this toxic concentration downstream to River Mile 20.

Copper was found at concentrations equal to or greater than its chronic toxicity criteria throughout the river (Figure 3.74). The maximum concentration measured

Table 3.17   Major Wastewater Discharges to the Blackstone River, 1993

Name	Town	Receiving stream	Flow in cfs	Suspended solids in ppm
Worcester Spinning	Leicester	Kettle Brook		
New England Plating	Worcester	Blackstone River		
Wyman Gordon	Worcester	Blackstone River		
Upper Blackstone District	Millbury	Blackstone River	56	9.2
Millbury Town	Millbury	Blackstone River	1.4	16.0
Wyman Gordon	Grafton	Blackstone River		
Grafton Town	Grafton	Blackstone River	1.8	5.4
COZ Chemical	Northbridge	Blackstone River		
Northbridge Town	Northbridge	Blackstone River		5.8
Douglas Town	Douglas	Mumford River	1.7	
Guilford, Inc.	Douglas	Mumford River		
Upton Town	Upton	West River	0.3	10.7
Uxbridge Town	Uxbridge	Blackstone River	0.9	12.8
Burrillville Town	Burrillville	Clear River	1.1	7.1
Hopedale Town	Hopedale	Mill River	0.6	7.5
Woonsocket Town	Woonsocket	Blackstone River	11.6	33.6
Okonite, Inc.	Ashton	Blackstone River		
GTE	Central Falls	Blackstone River		

was 40 ppb, compared to the EPA chronic criteria of 6.5. The concentration of lead was also higher than the EPA criteria of 1.3 ppb throughout the river. The combined effect of the three heavy metals on the microfauna in the river must have been severe.

Significant toxicity of river water to the water sow bug *Hyalella azteca* was observed in late October, but not in July when the river flow was higher and provided more dilution (Table 3.18). However, the overall toxic effect was reflected in the low richness index of the biota found on the stream bottom at River Miles 34 through 20 and at the mouth of the river (Figure 3.75).

3.4.6.4 Biological Richness

Despite the toxic concentration of heavy metals in the upper portions of the river, the richness index, determined for genus levels, was highest at station 39.7, exceeding 10.

In comparison with the Ten Mile River, however, the Blackstone River richness index was severely limited, less than one half the maximum value of 24, dropping to one half the minimum value of 10 noted on the Ten Mile River.

One explanation for the erratic toxic impact on the smaller aquatic organisms may be that the heavy metal toxicity is strongly affected by seasonal conditions, spring and summer being the times when the toxicity is blocked by sulfides in the sediments. In the winter, or after large storms, the sulfides release the metals, allowing them to exert the toxicity expected from laboratory studies. Unfortunately, most large river surveys were performed in the summer in the past, due to the more favorable conditions for working in the water. For this reason, much of the data on metal toxicity from field surveys underestimates the true toxic effects. It is now

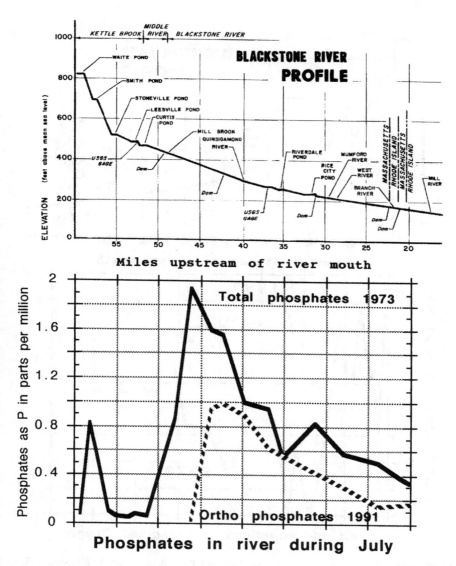

Figure 3.71 Phosphate nutrient concentrations in Blackstone River, 1973–1991.

understood that such river surveys for toxicity of heavy metals must also be performed in the winter.

3.4.7 Narragansett Bay

The four rivers that discharge into Narragansett Bay and eventually Rhode Island Sound have quite different characteristics, and require different strategies to develop sustainable ecologies. The Taunton River has the greatest potential for improving

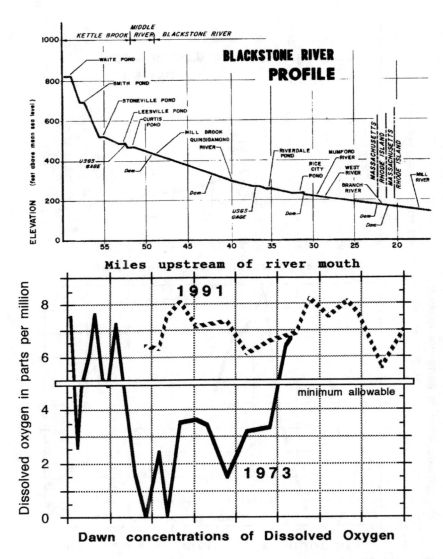

Figure 3.72 Minimum concentrations of dissolved oxygen observed in Blackstone River, 1973–1991.

fish production. It only has one dam on the main river, 41 miles upstream from the mouth, at the confluence of the Town and Matfield Rivers. Thus, migratory fish can make use of most of the river and tributaries over the first 41 miles. Also, heavy metals and other toxic materials are minimal.

In contrast, the other rivers (especially the Blackstone River) are blocked by dozens of small dams, and the rivers and their sediments are heavily spiced with toxic metals and other industrial chemicals inimical to aquatic life, especially to fish and to the organisms they require for food. Also, the long detention time in the many impoundments

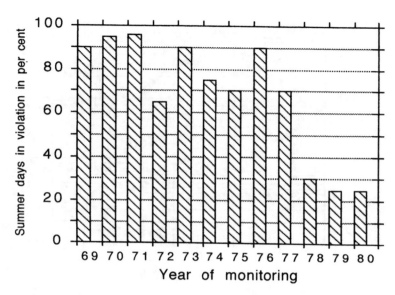

Figure 3.73 Percentage of summer days when dissolved oxygen concentration was in violation of the minimum oxygen standard in Blackstone River at the Massachusetts–Rhode Island state line. The new treatment plant for the Blackstone Valley Water Pollution Abatement District began to operate 21 miles upstream in 1978, showing a clear impact. (From Isaac, R. 1991. *Water Environment and Technology,* 25, 69–72. With permission.)

results in much higher water temperatures than desirable. Thus, the best efforts of state and federal agencies during the Green Movement have not improved conditions enough in these rivers to make them fishable, or even swimmable.

Contamination of the Ten Mile River has been so brutal in the last century that one wonders if it can ever be restored to its primeval quality. With most of the summer flow coming out of domestic sewage treatment plants, there is little chance that phosphate nutrients can be reduced to the necessary lower limit to avoid excessive algae and weed growth. Heavy metals are ubiquitous; they are not completely removed from the domestic sewage, even by advanced treatment. Deposits of these heavy metals remain in the sediments behind the many small dams on the main course of the Ten Mile River, and even on its tributaries. These metals may bind with sulfides or organic matter, but in combination with the dissolved and particulate matter that has long ago passed into Narragansett Bay, they represent a tremendous load of potential toxicants that have made their way into the estuary and coastal portions of Rhode Island Sound.

In addition to the industrial chemicals contaminating the Narragansett Bay system, a large load of lead from gasoline additives apparently poured into the river from the extensive roadways and interstate highways that parallel the river's course. Lead, copper, zinc, grease and oil, and combustion products are the documented contaminants coming from transportation systems. While the load of lead may have been curtailed in recent years, the other metals and hydrocarbons continue to wash into the river. Their ultimate resting place is the bottom of Narragansett Bay, their final location depending on hurricanes and strong currents.

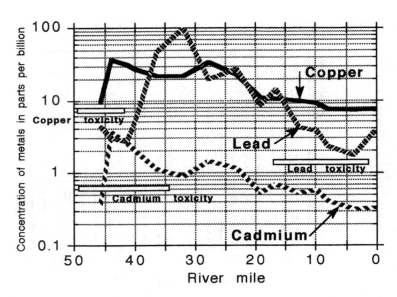

Figure 3.74   Cadmium, copper, and lead in Blackstone River, 1991. These toxic metals exceeded the concentrations causing chronic toxicity in aquatic organisms. Boxes indicate chronic toxicity values according to EPA criteria for the hardness encountered in the river.

Table 3.18   Survival of *Hyalella azteca* on Sediments from Impoundments in the Blackstone River, Summer and Autumn of 1991

Impoundment	July 22	September 22	October 30
Singing Dam	15%		2%
Fisherville Dam	70%		0
Rockdale Pond	86%		0
Rice City Pond	7%		0
Tupperware Dam		18%	13%
Mannville Dam		33%	2%
Slater Mill Dam		58%	25%
Lexington clean water reference	85%	90%	84%

Dealing with these toxic residues requires more than changes in discharge permits, and cannot be solved by construction of new sewage plants. Other possible approaches include:

1. Removing dams to reduce infiltration to sanitary sewers and consequent hydraulic overloading of treatment plants, and also reduce local flooding.
2. Low head hydropower can be used below steady flows of municipal sewage discharges.
3. Effluents from Attleboro and North Attleborough might be used for summer irrigation of lands downstream.

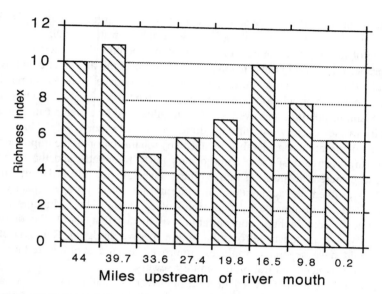

Figure 3.75 Genus-level richness indices for biotic communities on the bottom of Blackstone River, 1991.

In the Ten Mile River, with extremely low summer flows and a high proportion of wastewater discharges, contaminants in these discharges must be further restricted to protect the aquatic life, necessitating the issuance of more stringent discharge permits. The primary pollutants to consider in the discharge permits for both industrial and municipal discharges are copper, lead, and nickel. Ammonia and phosphorus must also be reduced from some discharges. Local effects from chlorine in effluents should also be corrected, and post-aeration of the effluent from the North Attleborough and Attleboro treatment plants should be resumed.

Industries can meet the new limits on contaminants by source reduction in their process streams, by additional treatment, or by connections to the municipal sewers after adequate pretreatment of their wastewaters. The town of North Attleborough should also begin a program of source reduction for copper in its drinking water system, perhaps to include additional adjustments to the pH and reduction of copper piping. Attleboro should soon complete its control of improperly managed industrial discharges presently connected to the municipal sewer system. In addition, both municipal treatment plants will have to improve their phosphate removal systems, preferably in a manner that will also enhance metal removal.

Improvements initiated through the 1984 round of discharge permits will not completely restore the river, particularly with respect to the downstream impoundments. Reduction of copper in the river will probably result in increased algal populations in these impoundments. Thus, problems with nutrients and algae dynamics should again be evaluated in the river after the metals have been reduced to acceptable levels. The possibility of nutrient dilution by augmentation of low summer flows should be investigated by evaluating existing and potential reservoir sites in the upper reaches of the drainage basin for storage of clean flows from winter and spring storms. This kind of reservoir site might also be useful for increased drinking water supply.

Some of the impoundments along the river are potentially attractive resources. Their proper rehabilitation and management could become an esthetic and even an economic asset to local communities. This is especially true in the lower reaches of the river where fish populations are returning, and where recreational use has already begun. Fishing, boating, and even swimming sites could be created in convenient proximity to the several communities along the river. Property values in Attleboro and Seekonk might be considerably enhanced by rehabilitation of the impoundments.

Flooding upstream of these dams occurs fairly frequently and the high water tables cause increased filtration into sanitary sewers, especially in North Attleborough. This problem is exaggerated when the normal flood storage capacity of the impoundments is eliminated by deposition of sediment. The sediments in these impoundments contain high concentrations of metals that should be detoxified, probably by state or federal efforts that might involve dredging. Removal of sediments would have several advantages: improved aquatic ecology, reduced local flooding, and reduced infiltration to sanitary sewers.

Although there is little precedent for it in a river as small as the Ten Mile River, formation of a river district or commission might be useful to coordinate the treatment of wastewater discharges and to evaluate and manage the numerous impoundments located in the several municipalities and two states. The following ideas for innovative approaches to correction of the river's problems deserve more thorough exploration if the river is ever to be truly restored.

1. Revenue for rehabilitation might be obtained by installation of hydroelectric generators in the higher dams, especially those downstream of the major discharges, which would provide steady flows.
2. The hydraulic energy available in the impoundments might also be utilized for power to effect local improvements in the river such as aeration, sediment removal, and even removal of residual color and contaminants from the main river flow. Removal of color would be a considerable improvement esthetically.
3. Mapping of sediments and analysis of the dam structures would be a necessary first step in developing such a river basin plan. A Ten Mile River Commission could conduct cost-effectiveness analyses in regard to reduction of algae and vegetation in the impoundments and thus improvement of the river for fish, comparing dredging of the sediments vs. increased chemical removal of phosphates in the treatment plants. Detoxification and dispersal of sediment from the impoundments could also be best evaluated with regard to the entire river basin, evaluating new provisions in the federal Superfund for hazardous wastes, or special legislative action. Considerable capacity for sludge incineration and land disposal of residues exists at the Attleboro treatment plant. This capacity might also be used for other treatment plants in the basin and perhaps even for disposal of some of the sediments in the river impoundments.
4. Careful consideration should be given to extraction of the large amounts of metal, including copper, silver, and gold, that have accumulated in the deeper sediments during the early part of the 1900s.

5. Control of contaminants from roadway and urban runoff throughout the river basin should also be initiated to minimize background levels of contaminants in the river, especially lead and phosphates. The unusual density of interstate highways in the river basin area, with four major roads and three interchanges, may be responsible for the widespread and increasing concentrations of lead, both in the water and in the flesh of fish. These interstate highways were completed in 1990s. Open land in the cloverleaf intersections of the interstate highways might be investigated for passive treatment of contaminants in the roadway runoff, and even for augmentation of low river flows.

3.5 LONG ISLAND SOUND

Long Island Sound is a narrow and sheltered, shallow portion of the North Atlantic Ocean off the coast of Connecticut, receiving freshwater flows from the Connecticut, Housatonic, and Thames Rivers. Most of the fresh water coming into Long Island Sound comes from the Connecticut and Housatonic Rivers, which drain western New England from as far north as the Canadian border. The Connecticut coastline was settled early in the colonial period of America because of the hundreds of protected bays and inlets, and the proximity of the protective barrier formed by Long Island to the south.

3.5.1 Connecticut River

In the Algonkian language, Connecticut means "long river." The Connecticut River begins in a series of small lakes slightly south of the U.S.–Canadian border, in northern New Hampshire (Figure 3.76). The northern-most limit of the watershed defines the corner of New Hampshire, abutting the Province of Quebec and the state of Maine. The longest river and largest river basin in New England, the Connecticut River flows for over 400 miles to Long Island Sound.

As the river flows south from Canada, it defines the border between Vermont and New Hampshire, passing between the Green Mountains on the west and the White Mountains on the east. The Connecticut River passes Hanover, New Hampshire, and Brattleboro, Vermont. The river then enters Massachusetts near Northfield, flowing into the state of Connecticut near Thompsonville, and finally discharging into Long Island Sound at Saybrook Point, opposite Greenport on the northeast corner of Long Island.

Although there have been a few interstate surveys of the entire Connecticut River, most of the data presented here is from the portion of the river in Massachusetts, including several of the major tributaries: the Millers, Deerfield, Westfield, and Chicopee Rivers. Although Quabbin Reservoir is in the geographical basin of the Connecticut River, the flow from the reservoir is diverted to metropolitan Boston on the East Coast, considerably reducing the contribution of the Chicopee River to the Connecticut River flow.

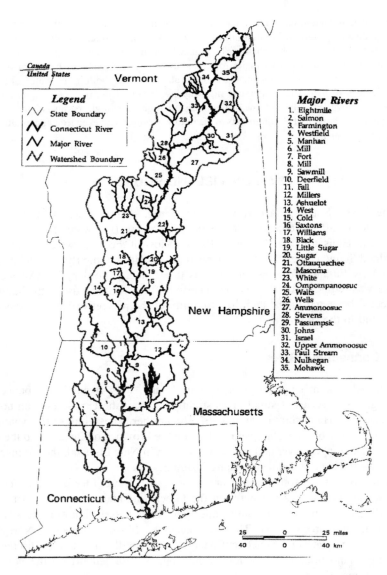

Figure 3.76   The Connecticut River is the longest river in New England. (From U.S. Fish and Wildlife Service, 1995.)

3.5.1.1 Historical Background

The Connecticut River system served as a waterway for travelers for centuries, both north and south on the mainstem, and east to west on the Chicopee River and other tributaries. The many rapids, narrows, and especially the waterfalls have always been important as fishing sites for the early Americans and for the immigrants during the colonial period.

Locations of important fishing sites can be traced through the names of towns along the rivers: Turners Falls, Chicopee Falls, Millers Falls, and Shelburne Falls. Holyoke was also a site of major waterfalls and fishing points for migratory fish such as shad and salmon.

As soon as the colonial communities became established, however, these same sites were developed for water power by building dams, weirs, and diversion channels. Consequently, the passage of migratory fish was impeded and often completely blocked, initiating a decline in the numbers of this important food source almost from the first days of the colonial period.

The mainstem of the Connecticut River and also some of its tributaries are bordered by wide floodplains of sediments deposited during the frequent floods. The floodplains continually erode and reform in a dynamic process that leaves remnant channels and oxbows as the main channel moves laterally.

These floodplain sediments form some of the finest agricultural lands in New England and have been used for that purpose for millennia. The most extensive floodplains in Massachusetts are near Agawam, West Springfield, Northampton, Hadley, Deerfield, and Northfield.

3.5.1.2 Restoration of Atlantic Salmon

Atlantic salmon were once abundant in the Connecticut River, ascending nearly 400 miles up the river to Beechers Falls in Vermont. The precolonial salmon population was enormous, supplying the original Americans and even the early colonists with an abundant supply of food. But the colonists soon turned industrialists. In 1798, the first dam across the main course of the river was constructed at Turners Falls, Massachusetts, preventing passage of the salmon. After their elimination a few years later, several futile attempts were made to restore the salmon runs.

In 1967, a more successful attempt began, with assistance from the U.S. Fish and Wildlife Service starting in 1979. However, progress has been extremely slow; the returning salmon number only in the hundreds at the end of the century.

Nonetheless, some improvement has occurred, and some basic information has been established, such as the global path of salmon migration (Figure 3.77). Salmon that hatch in the small tributaries of the Connecticut River, travel as far North as Greenland before returning to the tributary of their origin.

3.5.1.3 Restoration of Atlantic Shad

In contrast with the paltry returns of salmon, the Atlantic shad have returned to the Connecticut River in recent years in great numbers. Over 300,000 returned in 1995. Almost 200,000 passed the fish lift at Holyoke, Massachusetts; 18,000 made it up to Turners Falls in northern Massachusetts; and 147 even passed above Bellows Falls in southern Vermont.

This run in 1995 still does not approach the precolonial runs that may have numbered several million, but the progress is encouraging. In addition to shad, the migration of Blueback herring, sea lampreys, and striped bass have also been

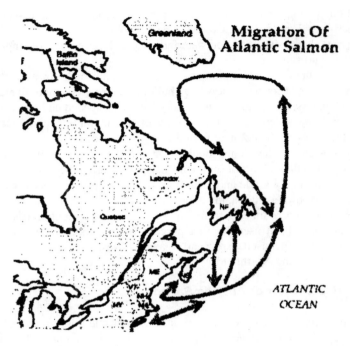

Figure 3.77 Atlantic salmon migrate from the Connecticut River as far north as Greenland. (From U.S. Fish and Wildlife Service, 1995.)

improved due to reductions in water pollution and provision of fish passage facilities at the numerous dams in the basin.

3.5.2 Housatonic River

The Housatonic River forms in Pittsfield, Massachusetts, and flows south through western Massachusetts for about 50 river miles, passing through the low but scenic Berkshire Mountains. The river glides past Stockbridge Bowl in Lee, Massachusetts, where the summer Berkshire Music Festival is conducted by the Boston Symphony Orchestra (Figure 3.78). This festival was a favorite of the modern American composer Leonard Bernstein, and is a favorite summer celebration for all New England as well as residents of New York.

The Housatonic River crosses into Connecticut about 1 mile north of Canaan, Connecticut, then continues for 85 idyllic and tranquil miles southward through western Connecticut and into Long Island Sound to the east of Bridgeport. The drainage basin includes 500 square miles in Massachusetts, 200 miles of eastern New York state, and 1230 miles of Connecticut.

Although the gradient of the headwater streams upstream of Pittsfield is quite steep, dropping from an elevation of 1500 feet above sea level to an elevation of 1000 feet in a distance of 10 miles, the rest of the river slope is more gradual, dropping only 340 feet in the next 55 miles to the Connecticut state line (Figure 3.79). The 85 miles through Connecticut are also fairly flat, a drop of 650 feet to the sea.

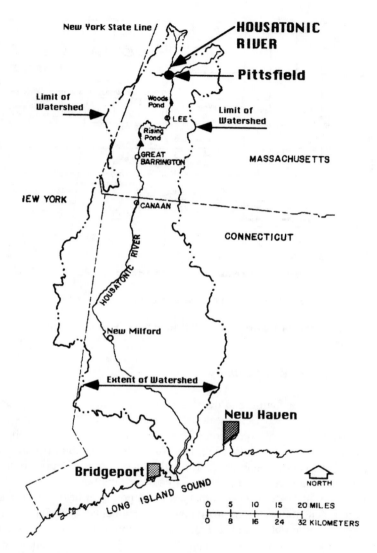

Figure 3.78 Housatonic River Basin in Massachusetts, Connecticut, and New York states, flowing into Long Island Sound.

Above Pittsfield, the energy of the East Branch is captured by eight small dams, followed by seven more dams in Massachusetts, and several more in Connecticut. The 15 dams in Massachusetts were constructed in the 19th Century by paper industries that formed the basis of the local economy.

From the early 1930s the General Electric Company established a facility in Pittsfield to manufacture electrical transformers. At the close of the 20th Century, the manufacturing of electrical machinery and the production of paper continued to be the major activities in the local economy, supplemented by tourism and recreation.

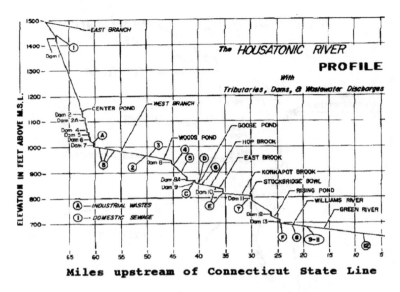

Figure 3.79 Profile of upper Housatonic River in Massachusetts.

The picturesque towns of Lenox and Stockbridge have been major tourist attractions for many years.

3.5.2.1 History of Water Quality

The increase in population and development of industry along the Housatonic River during the early 19th Century destroyed the purity of the river. As early as 1884, local writers noted discoloration of the water by mill wastes, a problem that continues even a century later. In 1936, the Works Progress Administration of the national government documented extensive pollution of the river, and recommended the construction of domestic and industrial treatment facilities.

A state survey in 1949 again documented gross pollution of the river, resulting in more recommendations for improvement. Governmental agencies continually studied the pollution problem and recommended action, usually involving creation of another agency. The New England–New York State Interagency Committee published another report on the river in 1954, adding the concerns of flood control and drainage, navigation and power production, fisheries, wildlife, and recreation to the initial concerns for water quality. This was followed by additional reports by federal agencies, and then a thorough study of water quality and sources of pollution in 1969 (Table 3.19).

The 1969 study by the Massachusetts Division of Water Pollution Control was preceded by formal classification of the upper river for boating and fishing, and for swimming below Great Barrington and in the tributary streams (Figure 3.78). It was determined that wastewater discharges from the city of Pittsfield and the General Electric Company caused the most severe depletion of oxygen of all discharges to

Table 3.19   Domestic and Industrial Wastewater Sources in Housatonic River, 1969

Source of wastewater[a]	Industry or municipality	River mile upstream of Connecticut state line
Industry		
A	Crane Paper of Dalton	60.9
B	General Electric of Pittsfield	59.3, 57.1, 56.9
C	Schweitzer of Lee	42
D	Westfield River Paper of Lee	40
E	Hurlbut of Lee	36.5, 35.3
F	Rising Paper of Great Barrington	24.4
Municipality		
1	Hinsdale	67.4
2	Pittsfield and Dalton	50.9
3	North Lenox	49.1
4	Lenox Center	45
5	Lenoxdale	43.6
6	Lee	39.2
7	Stockbridge	30.5
8	West Stockbridge	23.3
9–11	Great Barrington	19.5, 25.7
12	Sheffield	8.5

[a] See Figure 3.79 to locate wastewater sources.

the river. The dissolved oxygen concentration fell below 2 parts per million (ppm) during July 1969 at Station B below the General Electric Company discharge, and did not recover to the required 5 ppm for 5 miles downstream (Figure 3.80). It was concluded that these discharges would require high degrees of treatment, including removal of nutrients. Dissolved oxygen concentrations were acceptable in Woods Pond at River Mile 46 and downstream in 1969.

The concentration of suspended solids reached 7 to 12 ppm as the river passed through Pittsfield at River Mile 55, and then soared to 16 and 19 ppm in Goose Pond at River Mile 40 (Figure 3.81). The concentrations of suspended solids remained high all the way to Connecticut — from 7 to 17 ppm.

Phosphate nutrient concentrations were extremely high throughout the river, and the river water was severely discolored below Woods Pond (Figures 3.82 and 3.83). The number of fecal coliform bacteria were also high throughout the river in Massachusetts during the summer of 1969.

After analysis of the 1969 field data, the state agency ordered the municipalities and industries to improve their treatment facilities, based on calculations developed with the aid of a new computer simulation of the river. The Massachusetts Division of Water Pollution Control had contracted with a consulting firm to develop a generalized computer model that could be used for predicting oxygen concentrations in all rivers of Massachusetts, and had used the Housatonic River as the first trial of the model.

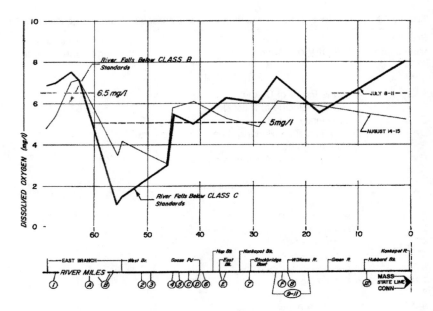

Figure 3.80　Dissolved oxygen concentrations in upper Housatonic River, 1969. See Figure 3.79 and Table 3.19 for keys to bottom legend.

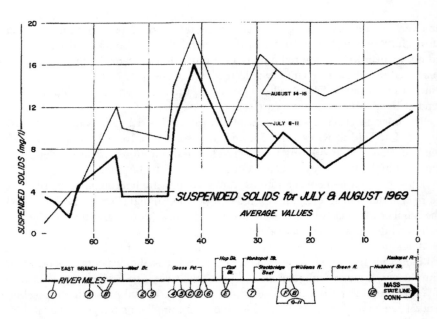

Figure 3.81　Suspended solids concentrations in upper Housatonic River, 1969. See Figure 3.79 and Table 3.19 for keys to bottom legend.

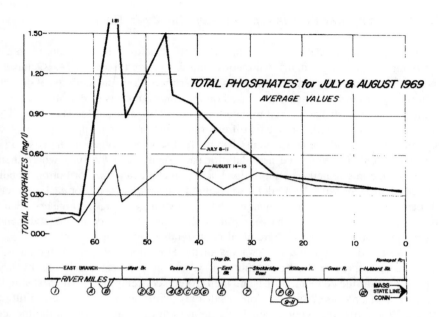

Figure 3.82   Total phosphate nutrient concentrations in upper Housatonic River, 1969. See
Figure 3.79 and Table 3.19 for keys to bottom legend.

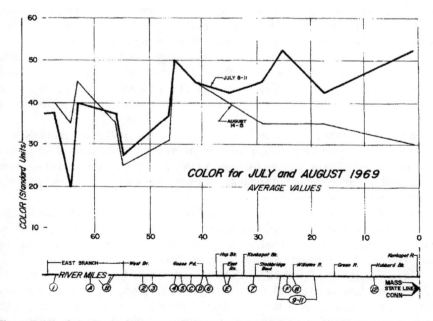

Figure 3.83   Color in upper Housatonic River, 1969. See Figure 3.79 and Table 3.19 for keys
to bottom legend.

3.5.2.2   Computer Simulation of Water Quality in Rivers

The computer model developed in 1969 was structured in segments to reflect the points of discharge, dams, tributaries, and reservoirs typical of Massachusetts rivers, and was used for over two decades to determine allowable loadings of organic material in wastewater discharges, based on the concentrations of dissolved oxygen required to protect fisheries, and to determine allowable nutrient loads as well. The computer model included a set of differential equations for computing decay of organic material and the subsequent changes in dissolved oxygen in the river, in a repeated calculation that used new data for each segment of the river where there was a change in geometry, pollution load, aeration characteristics, or bottom conditions. Calculations with the model also required the time of travel for each segment of the river as a function of the river discharge, as well as several parameters related to oxygen depletion or aeration over rapids or dam spillways, over organic sediments and through impoundments where algal photosynthesis was significant.

Using the data from the 1969 survey of the Housatonic River, the model was calibrated to represent the true salinity, flow, and geometrical conditions by first simulating chloride concentrations, and then oxygen concentrations. The first simple simulation of chloride concentrations dealt only with the basic flow and dilution effects. When this was accomplished with minor adjustments to data in the model, it was concluded that the basic equations and structure of the computer program were correct. Then, the more complicated simulation of dissolved oxygen was conducted. The model was easily adjusted to simulate the observed oxygen concentrations throughout the river, again indicating that the computer program was adequate.

The calibrated model was then used to evaluate proposed flows and wastewater loads for the various industries and municipalities on the upper portion of the river. The variables considered in the proposed discharges were the flow rate and the concentrations of organic material, oxygen, and phosphates. Simulations were conducted for a variety of geographical and temporal arrangements in the wastewater discharges, and for drought conditions in the river that usually occur in the late summer.

In the simulation process, the concentrations of the polluting materials were reduced until the simulated oxygen concentrations in the river were always above the 5 ppm required to protect the fisheries. In this way, the allowable loadings were determined for organic material and nutrients in the wastewater discharges proposed by the industries and municipalities.

The industries and municipalities were then able to determine the amount and type of treatment needed for their wastewaters, in order not to exceed the allowable loadings calculated from the model. This procedure was a very careful and rational way to design the treatment facilities to protect fisheries in the river.

This computer model, named STREAM, was then used for the next three decades on all Massachusetts rivers, as new discharges were proposed or as existing discharges caused violations of the water quality standards in the river receiving the discharge. As each river was evaluated and reevaluated, the accuracy of the data for the river was improved, resulting in a highly reliable model for allocation of waste-water loads. Simulations of Boston Harbor and other coastal bodies of water were also developed, based on the same differential equations for transport of chlorides and oxygen, but set in a two-dimensional framework, instead of the single-dimensional structure of the river models.

Although precise concentrations of pollutants could be calculated for each waste-water discharge, it was not always easy for the polluters to reduce their waste loads to those concentrations. Unfortunately, the regulatory agencies often agreed to installation of the best available technology, rather than require the polluter to install very expensive treatment systems, or to radically change the process that produced the pollutants.

3.5.2.3 Water Quality at the Close of the Century

The most recent water quality surveys on the Housatonic River indicated that improvements immediately after 1969 were large, but further changes in the last two decades of the century were very small, barely able to keep pace with the increased waste loads from the growing human populations and industries. However, the problems with discoloration and low oxygen had been corrected, based on the most recent survey of 1992. Dissolved oxygen concentrations throughout the summer of 1992 were also satisfactory, always above 6 ppm. Thus, they exceeded the minimum requirement of 5 ppm for the mean concentration of dissolved oxygen.

Unfortunately, the stubborn problems of excessive phosphate nutrients and high suspended solids concentrations were not under control at the end of the century, despite 3 decades of aggressive and intensive programs to control pollution sources in the basin. The concentration of total phosphates decreased to roughly 0.1 ppm by 1992, significantly below the values for 1985 and markedly lower than the values in 1967, but still twice the desirable limit for control of algae growth in impoundments (Figure 3.84).

In 1992, the highest concentration of total phosphates was 0.16 ppm, just below the discharge from the city of Pittsfield treatment facility (Figure 3.84). The large load of phosphates from Pittsfield caused elevated concentrations to persist for the next 50 miles. The phosphate concentrations were twice the desirable value for about 10 miles downstream of the Pittsfield discharge, stimulating algae growths in the several impoundments, including Woods Pond at River Mile 46.

The algae populations stimulated by the excessive phosphates caused high values in the measurements of suspended solids in 1992. The concentration of suspended solids was about 8 ppm as the river passed through the six small impoundments in the middle 30 miles of the river in Massachusetts, then rose to 15 ppm at mile 25 below the discharge from Rising Paper in Great Barrington (Figure 3.85). Downstream, the suspended solids decreased to nearly normal values. The only reach of river that had lower suspended solids in 1992 was the last 10 miles upstream of the Connecticut state line. Because of the stimulated algae populations in Housatonic River impoundments, the suspended solids concentration in the river in 1992 were not markedly lower than those recorded 7 years earlier in 1985 (Figure 3.85).

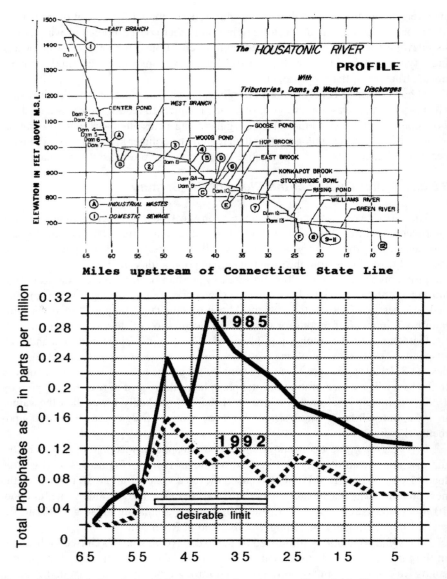

Figure 3.84 Total phosphate nutrient concentrations in upper Housatonic River, 1985–1992.

3.5.2.4 *Polychlorinated Biphenyls*

From 1934 to 1977, polychlorinated biphenyl compounds (PCBs) were used by the General Electric Company in Pittsfield in the manufacture of electrical transformers, and a great deal of this toxic chemical leaked into the surrounding environment, including the Housatonic River. In 1977, recognition of the toxicity and environmental persistence of the polychlorinated biphenyls resulted in a nationwide

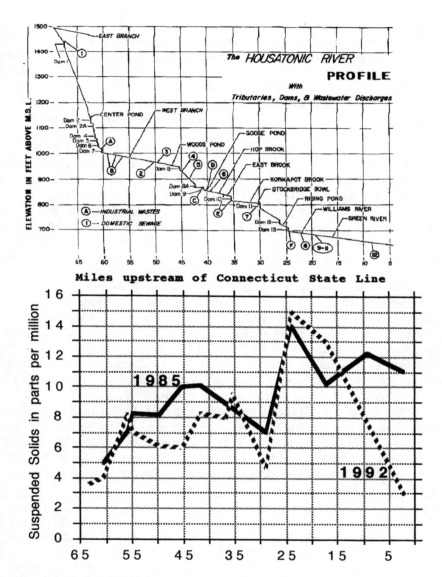

Figure 3.85 Suspended solids concentrations in upper Housatonic River, 1985–1992.

ban on PCBs except in totally enclosed systems, under provisions of the federal Toxic Substances Control Act of 1976.

PCBs had been detected in sediments of the river and its impoundments in Connecticut from 1974 to 1977. Fish in the river in Connecticut were found to have more than the desirable 5 ppm of PCBs in their flesh in 1977, and fish in Massachusetts had even higher concentrations. The high concentrations of PCBs in the river sediments caused concern about contamination of the underlying aquifers that supplied many of the riverside communities with drinking water.

Surveys along the entire river in Massachusetts in 1981 showed that PCBs were in all of the river sediments downstream of Pittsfield, primarily in impoundments caused by the several small dams. However, the highest concentrations, reaching 140 ppm, were found in the river sediments between Pittsfield and Woods Pond in Lee. About half of all the contaminants in the river were found in Woods Pond — approximately 11,000 pounds of PCBs. The contaminated sediments were gradually being buried under newer, cleaner sediments arriving since the elimination of the use of PCBs by General Electric in 1977.

Although careful studies indicated that the PCBs were not penetrating the aquifers underlying the impoundments on the river, the resuspended contaminant was being slowly washed over the dams, and there was considerable danger of a large release in the event of dam failure or a large storm. Even in a normal year, about 500 pounds of PCBs were being transported as sediments resuspended by the river from these impoundments. The suspended contaminants flowed into Connecticut and eventually to Long Island Sound where they mixed and probably settled in coastal waters, habitat for many fish, shellfish, and crustacea used for human food. It was also concluded that low concentrations of dissolved PCBs were being continuously carried by the river.

The future risks from spread of this highly toxic chemical from its encapsulated deposit behind the dams along the Housatonic River to aquatic and marine organisms downstream, typifies a widespread danger from industrial contaminants in reservoir sediments. Persistent and toxic materials such as the PCBs, heavy metals, and other synthetic compounds are temporarily arrested by the dams, but may suddenly be recirculated in the event of hurricanes, large floods, or failure of the dams. As many of these dams are gradually being abandoned, the risk may occur at a time when it will be difficult to determine liability or require remediation.

Analysis of flow records and PCB transport at the U.S. Geological Survey gaging station at Great Barrington indicated that slightly less than 1 part per billion (ppb) PCB was transported for flood flows below 3000 cubic feet per second (cfs). However, the 20-year flood was 8000 cfs, and floods above 10,000 cfs occur about every 50 years, probably sufficient to cause considerable transport of sediments and contaminants.

It worries me that, although one extremely contaminated portion of land near General Electric in Pittsfield has been placed under the Superfund process, the contaminated reservoirs in the river are not being addressed by state or federal programs. Recent evidence shows that the PCBs are spread all over the landscape as well. This is an important and dangerous problem that needs to be addressed. Recent studies on sturgeon larvae in the Connecticut River showed that similar chemicals in river sediment, such as coal tar, were still highly toxic after being buried for 40 years.

3.6 SUMMARY

Important discoveries in our historic review of the rivers and harbors of New England can be summarized in terms of each of the individual rivers, and of the major bays.

From the Neponset River leading to Boston Harbor, we learned that water pollution by industrial corporations has caused significant property damage, and that these corporations avoid responsibility for the damage through their collective manipulation of environmental agencies and regulations. However, the depression in property values due to water pollution may be an opportunity for citizen groups to buy up river and lake shores now, for the use of future generations when the rivers are restored.

Field studies in Foxboro by student activists on the toxic impact of industrial contaminants such as heavy metals, demonstrated the need to monitor water quality during all four seasons in New England rivers and lakes, and the need to closely track the downstream repercussions of large storms and hurricanes on the quality of these waters.

From the Taunton River in southeastern Massachusetts, we learned that there are many opportunities to use treated sewage for crop irrigation and sewage sludge for agricultural purposes. This depends on local geography and climate, but it is especially true where river flows become extremely low in the summer, and water quality considerations would require excessive costs to give advanced treatment of sewage discharges to the nearly dry rivers.

We also saw the value in focusing on restoration of valuable species of migratory fish as a way of unifying the environmental activities on rivers in the future. This will be far more creative and productive than previous emphases on regulation of discharges and simple water quality criteria.

A review of environmental enforcement proceedings on the Ten Mile River showed the gradual growth in effectiveness of regulatory programs, and then the strongly negative reaction from the affected industries that led to the recent death of the Green Movement. The Greens won the battles, but lost the war.

The Ten Mile River was also a good example of the gradual decay and abandonment of the myriads of small mill dams on New England rivers. These dams no longer serve useful functions, and many are ecologically dangerous because of the toxic sediments they store and their physical obstruction of passage by migratory fish. A new approach to these old dams is needed.

The four centuries of environmental history of Boston Harbor and its cyclical course revealed the need for imaginative thinking by water engineers and planners. Concrete and steel solutions will no longer be acceptable for solving such ecological problems. It is time for water planners to include ecology, human behavior, and history in their deliberations.

Specifically in Boston Harbor, there is an approaching crisis when the authorities will have to use their collective wisdom, instead of blindly following federal regulations. One can hope that New Englanders can again lead the nation in another revolution — this time, an environmental revolution.

3.6.1   A Comparison of Narragansett and Massachusetts Bays

There is some value in comparing the dams and ecology of the two major bays in New England: Narragansett Bay and Massachusetts Bay. Both drain the same low hills of central Massachusetts and both open into the North Atlantic near the great fisheries of Georges Bank. And the rivers that feed both bays have been blocked by myriad small dams, built for power and industry at the turn of the century. But there the similarities end.

The rate of flow of fresh water through Boston Harbor and into Massachusetts Bay has increased over past centuries because of diversions from dams on rivers of central Massachusetts that formed Quabbin and Wachusett Reservoirs. The flow was then diverted out of those river valleys and into water supply and sewerage systems that eventually drained into Boston Harbor. This pristine and dependable flow of water was used by metropolitan dwellers and then discharged to the North and South Drainage systems of the MDC and the MWRA.

At present, this extra flow reaches the rivers and shores of the harbor at several points, adding to the natural process of harbor flushing. However, by the end of the century, all of this diverted water, plus considerable additional flow extracted from the local aquifers, will be discharged 10 miles offshore, after passing through the large, new central treatment plant on Deer Island at the mouth of Boston Harbor. The treated sewage will surface in the Atlantic halfway to the whale habitats at Stellwagen Bank (Figure 3.86). Unfortunately, this flow will no longer increase the harbor flushing rate. The MWRA discharge is significant; in 1996, it exceeded 500 cfs.

In contrast with Massachusetts Bay, the sewerage of Narragansett Bay has developed as a diffuse set of pipes and treatment plants throughout the valleys of the four rivers that feed the bay. The main sewage treatment plants for Providence, metropolitan Rhode Island, and the drainage system of Mt. Hope Bay have never been conglomerated into a single system. All of the discharges go directly to the rivers; none are displaced to the open ocean, as in Massachusetts Bay.

The comparative impacts of the concentrated discharge of Massachusetts Bay vs. the diffuse discharges of Narragansett Bay have a large impact on transport of organic matter to the bay systems, and the food supply for fish and whales that use the bays. In simple terms, the organic material discharged from the ocean diffusers to be constructed in Massachusetts Bay will eventually feed the whales and pelagic fish, while the material coming into Narragansett Bay is gradually released into the rivers, thus feeding the coastal fish and harbor organisms.

Narragansett Bay is more open to tidal exchange than is Boston Harbor, but has suffered more from hurricanes because of its exposure to the southern latitudes. The great hurricane of 1938 drove straight up Narragansett Bay, causing havoc and flooding through Providence and into Massachusetts. More recent hurricanes in 1955, 1984, and 1991 have also bruised Rhode Island, usually passing over Narragansett Bay.

The many communities around the bay have constructed treatment plants for their domestic and industrial wastewaters during the last half of the 20th Century, but over 10 million pounds of suspended solids continue to flow into the bay from these facilities, and the amount is again rising, since 1992 (Figure 3.87). The reversal in the previous downward trend in loadings is symptomatic of the general problems in the long-scale cycles of environmental improvements. As the flows increase, municipalities are less willing to bear the expanding costs of operation and maintenance, and treatment efficiency starts to decrease, more so as the treatment facilities age.

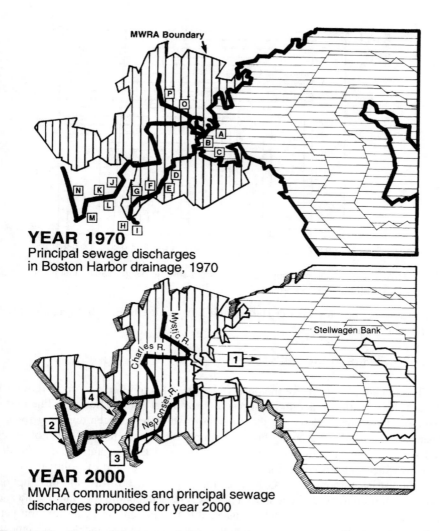

YEAR 1970
Principal sewage discharges
in Boston Harbor drainage, 1970

YEAR 2000
MWRA communities and principal sewage
discharges proposed for year 2000

Figure 3.86 Changes in location of major sewage discharges to Massachusetts Bay Drainage Basin, 1970–2000. Discharge points shown from all agencies in 1970 (letters A–P in upper diagram). Discharge point shown from MWRA sewage system in the year 2000 (number 1 in lower diagram), and from others in upper basin (numbers 2–4 in lower diagram).

Environmental agencies in Rhode Island were severely curtailed at the end of the 1990s, due to budget cuts in both federal and state government. This is a rather universal trend in the U.S., marking the end of the Green Revolution. Here is where the difference in the two bays will show its greatest impact. As sewage treatment fails in Narragansett Bay, the harbor will again become a greasy cesspool. But in Massachusetts Bay, the main discharge will be at sea, and conditions in the harbor will gradually improve as long as the sewer pipes and tide gates are maintained.

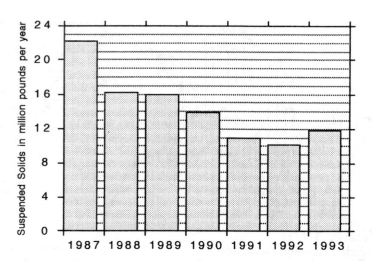

Figure 3.87   Annual loads of suspended solids from sewage discharges into all tributaries of Narragansett Bay Drainage Basin, 1987–1993.

Thus, there is hope for Boston Harbor and Massachusetts Bay, but tragedy ahead for Narragansett Bay.

3.6.1.1   *Cycles of the Bays*

The fundamental basis for the observed cycles of activity in Boston Harbor over the last three centuries, and the observed improvement and predicted decline in the ecology of Narragansett Bay, is the continuous struggle between opposing forces. There are the needs of growing human communities for increasing flows of water and sewage, countered by the equally aggressive resistance against paying for the costs of water and sewage treatment by these communities and by politically powerful industrial corporations. As long as the population growth and consumption rates are allowed to expand uncontrolled, the cycles will continue — more rapid and more brutish each time. Only when sustainable populations and consumption patterns are developed can one hope to achieve a balanced and sustainable environment.

3.6.1.2   *Forest to Farms to Forests*

In New England, there has been spectacular success in one aspect of environmental stability: the return of the forest. Unfortunately, this has been at the expense of an important economic sector: the New England farm. It is easy to understand why the farms disappeared. Farming in rocky New England soil was always a laborious and unrewarding endeavor. The soil is thin, deposited by about 10,000 years of decaying oak leaves and pine needles on top of small patches of sand left by the glaciers. The soil is acid, again the product of oak leaves and granite, and now made more acidic by acid rain.

Weather in New England was also never in favor of the farmer. But by hard labor and persistence, many families were able to grow the corn, pumpkins, apples, and other crops that sustained them. The broad riverbeds were often good for grazing dairy cattle, which could be moved out of the way when floods came. Cranberries could also be grown in the sandy and acid bogs along small streams. The New England farmer had a reputation for rugged persistence in the face of all these obstacles. Much of the crusty character of New England fame was due to these farmers.

To start a new farm, it was necessary to chop down the mixed forests and then pull out the annual crop of rocks placed there in past millennia by the glaciers. Sharp axes and strong farm boys quickly took down the forests, converting them into houses and barns. By the middle of the 18th Century, the hills of New England were clear-cut, turned into cornfields, hay meadows, alfalfa fields, and pasture. However, along with the forests went the deer, wild turkey, bear, moose, and the entire assemblage of forest animals that had sustained the original Americans and the first colonists. They were easy to hunt and then exterminate as the fragmented forests shrank. Eventually, they were found only in the rockiest and most remote corners of northern Maine where farming could not survive the long winter seasons. The ecology of the forests and woodlands was destroyed. But not forever.

As the farms in fertile flood plains of the Midwest began to compete with the rocky and struggling New England farms after the Civil War, the new railroads made it too easy to ship corn and wheat back east, and the New England farms lost the competition. Slowly, the farmers went west or went to work in the new mills springing up along the coastal rivers. As one farm after another was abandoned or sold, the pine trees and other evergreen softwoods came back with lightning speed and, eventually, the slower but persistent oaks and other hardwoods also joined the resettlement effort. Mixed forests began to reappear all over the hills and undeveloped lots. By the end of the 20th Century, New England towns and rural areas were again approaching the woodland ecology of the 16th Century. Deer and moose were moving back from the north. Wild turkey were coming back to the forests. It was not exactly the same as before, but a return had begun.

The rivers and bays of New England would have shared in this recovery if it were not for the hundreds of small dams built by the industrious immigrants during the past centuries. Studies of forested and clear-cut areas along the Pacific coast have shown that the falling trees and branches along mountain streams are an important source of organic carbon and detritus for the downstream rivers and estuaries, and even for the open ocean. But if the transport of these fallen branches is blocked by dams, and if the wood is periodically cleared from the reservoirs or dam spillways, it never finds its way to the coastal waters. In New England, this has prevented the reestablishment of an important part of the coastal food chain for marine and freshwater organisms.

3.6.1.3 The Clean Water Act Was a Construction Grants Program

A critical look at progress made during the Green Movement toward restoration of rivers and harbors in New England reveals that it did not work as expected.

Although the goals of the Clean Water Act were framed in terms of making it possible to swim and fish again, the pursuit of these goals was organized as a Construction Grants Program.

Succumbing to the traditional "concrete fever" of previous sanitary and green awakenings, hundreds of pipelines and treatment plants were constructed. Quite often, they were designed to raise the oxygen concentrations in the rivers; thus, it appeared they were protecting the fish. However, the driving force behind these construction projects was the political excitement generated by the concrete mixers and backhoes. Once the construction project was completed, the fish were forgotten. Fingers were pointed with pride at the new technology, beautifully landscaped and painted. Every town and many industries completed their construction project and assumed that was the end of the dirty water problem.

The big construction projects financed by the Clean Water Act did improve one component needed to restore fishing and swimming. They generally reduced the amount of fecal contamination in the rivers, and improved the low oxygen problems in the summer months. But swimming and fishing require much more than this.

Establishment of primeval fisheries in the coastal rivers of New England also requires restoration of tolerable and low water temperatures. Stimulation of migration by fish requires adequate spring floods in the estuaries, removal of physical barriers such as dams, and removal of chemical barriers such as reservoirs full of heavy metals. Sufficient flows during the dry summer months are needed to maintain warm-water fisheries. None of these were addressed by the Construction Grants Program of the Clean Water Act. More importantly, an economic connection was never made between the needs of the fisheries and the people who would have to meet these needs. For example, the toxicity of heavy metals in sediments of industrial reservoirs was only addressed technically in the 1990s, with no effective mechanism established to make the responsible industries remove the toxicity.

3.6.1.4 Reactions Instead of Principles

There were several important lessons learned in the temporary successes of the Green Movement in decreasing the sewage discharges to New England rivers, and in the lack of successes in dealing with the industrial dams forming these reservoirs. The temporary successes in dealing with sewage discharges were due to a repetition of the behavior in all the other past awakenings, by simply reacting to unbearable conditions.

When rivers and harbors became so foul that the fish were diseased and dying, when lakes became coated with offensive layers of fecal material and algae, then construction projects were established to reduce the sewage loads on these desecrated waters, and the problems were abated until tolerable. The facade of science and technology used in these abatement projects gave the impression that lasting and fundamental corrections had been made to the problems. They had not.

Other problems discovered in the course of the field surveys of water pollution in New England rivers did not have the same initial success as the Construction

Grants Program. These fatherless projects include diffuse pollution from roadsides, farmland, and residential areas, known by twisted logic as the non-point pollution sources. Regulations have been developed to try to minimize these diffuse sources, but they all require expenditures by individual polluters without being susceptible to remediation with a large construction grant. Thus, the success of controlling diffuse pollution depended on strong enforcement programs, which are never popular. Laws and regulations prohibiting contamination of the environment have been formulated and passed for generations, but with little effect.

The problems created by the industrial dams on the coastal rivers show more clearly the reasons for failure of well-intentioned efforts to restore fishing and swimming to our surface waters. There was nothing that could be constructed to remedy the damage caused by the dams. And there is no excitement or sense of creativity in such a destructive process as removal of the dams; thus, it was never seriously considered.

Many of the larger environmental programs of the Green Movement eventually came down to whether one could convince the legislature that it would involve construction projects with glamor, jobs, and the pride of creation. If there was no "concrete fever," there would be no program. The legislators were accustomed to support from the civil engineers and the construction industries and labor unions, who would have direct and immediate payoffs from such projects.

The publicity about the Boston Harbor Project of the MWRA often boasts of it being the largest civil works project in New England, that it is providing large numbers of jobs for the region, and that it is stimulating the economy. This is the kind of logic needed to overcome the staggering costs involved, and it was usually successful in the early stages of promoting such large construction projects.

3.6.2 Restoration of the Waters of New England

A new approach is needed to restore the ecology of the waters of New England with a sustainable program, based on self-supporting and sustainable solutions. Individual projects are outlined below for the rivers that flow into Massachusetts Bay and into Narragansett Bay, and especially for Boston Harbor and the communities served by the Massachusetts Water Resources Authority (the MWRA).

Some of these solutions are obvious, some may seem risky, and some may seem very difficult. But they are based on the considerable data reviewed in previous chapters and should be taken very seriously.

In order to generate a degree of excitement about these restoration activities that might approach the impact of "concrete fever," we must adopt the approach that we can "save the salmon," or "save the whales," or "restore the herring." Living symbols need to be elevated for all to see as the focus for these efforts.

3.6.2.1 MWRA, Boston Harbor, and Whales in Massachusetts Bay

The sustainable solution to cleaning up Boston Harbor and providing water and sewage treatment for the MWRA area should proceed as follows:

- Complete the current construction program.
  - Complete the Deer Island plant with complete primary but minimal secondary facilities.
  - Complete the deep ocean outfall.
  - Complete the renovation of the entire sewerage system.
  - Reduce all the toxic compounds coming into the system until the sludge is satisfactory for ocean disposal, and for consumption by plankton and krill, which are food for fish, lobsters, clams, and whales.
- Limit treatment to primary facilities.
  Operate only primary treatment facilities unless water quality conditions in Massachusetts Bay are unsatisfactory. Primary treatment will be adequate as currently improved conditions in the harbor and the bay are occurring without secondary treatment. Thus, the MWRA should:
  - Operate only the primary treatment system and sludge digestors at the new Deer Island facility, without using the secondary treatment system, thus saving huge operational costs.
  - The harbor will be cleaned up as soon as the deep-ocean outfall is used, as the sewers around the harbor and tributary streams have already been repaired and renovated to acceptable condition.
  - Rely on ocean dispersion and diffusion to keep oxygen, clarity, nutrients, and pathogens at acceptable levels in Massachusetts Bay. Use secondary treatment at Deer Island only when concentrations of these parameters become excessive.
  - Nutrients (nitrogen, phosphorus, and potassium) and suspended organics (carbon) will fertilize Massachusetts Bay, providing enormous food supplies for plankton and krill, which feed pelagic fish and whales, thus speeding the recovery of their populations in coastal waters.
  - Sludge pellets from MWRA digestors should be dispersed in feeding areas for whales and important pelagic fish (cod, halibut, haddock) to further stimulate rapid recovery of these depleted populations.
- Develop a sustainable system.
  Develop sustainable, steady-state water and sewage systems for metropolitan Boston and surrounding communities so that one does not exhaust the freshwater resources of New England and repeat the cyclical mistakes of the past in contaminating Boston Harbor. Construction of the dams that formed Quabbin Reservoir, Wachusett Reservoir, and other facilities of the MWRA have caused substantial losses of farms, real estate, wildlife, and fisheries in those regions. This kind of expansion will not be possible in the future. Thus, the MWRA should:
  - Put a cap on additional flows and sewage loadings in the MWRA system, including water consumption and sewage discharges.
    - Reduce water consumption by requiring lawns to be watered with greywater from individual homes.
    - Reduce sewage flows and loads by requiring composting toilets on new homes.
    - Develop alternative sources and supply secondary networks of nonsanitary water from existing reservoirs, also for use in lawn watering, composting toilets, and street washing.

— Encourage the peripheral tier of MWRA communities to withdraw from MWRA by developing their own tertiary treatment systems for domestic sewage, and limiting their flows and loads by:
  ○ Using greywater for lawns and flushing toilets.
  ○ Using composting toilets.
  ○ Eliminating industrial and other toxic wastes in their systems, and using the sewage effluent in the summer to irrigate alfalfa and other crops, instead of discharging to the rivers during the low-flow periods of the hot summer.
  ○ On a town-wide basis, use river water and rainwater taken from existing industrial reservoirs to supply secondary network of nonsanitary water to homes and industries for watering lawns, for flushing toilets, and for other non-human uses.
• Economic benefits.
  The economic benefits of this sustainable program for Massachusetts Bay program will be:
— Restoration of fishing in waters of New England and other East Coast fishing banks.
— More rapid recovery of fish and whale populations on Stellwagen Bank, which are valuable now as tourist attractions, and eventually as harvestable resources (Figure 3.88).
— Enormous savings in operating costs for MWRA, and lower sewer rates for ratepayers.
— Elimination of the need for new reservoirs and aqueducts for the Boston water supply system, and new collection and treatment facilities for the Boston Harbor system.
— Increased coastal productivity through the use of sewage sludge to supplement carbon-based food supply for organisms in swamps, estuaries, and coastal habitats. This food supply used to come from spring and hurricane storm transport of fallen logs from the primeval forests.
— Reduce contamination of harbor sediments and disposal costs for dredged material.

3.6.2.2 Restoration of Neponset River Fisheries — Save the Herring

Because of strong citizen alliances on the Neponset, Charles Nashua, and other rivers of New England, there is an important opportunity for citizen direction of river restoration along these rivers. The Neponset River was selected in 1994 by the governor of Massachusetts for special attention, and this beginning should be continued by the active citizen groups along the river. The same activities can be undertaken on the other rivers by the appropriate river, lake, and watershed associations.

The program for the Neponset River should begin as follows:

• Establish Neponset River Fisheries and Wildlife Authority as a state entity or regional agreement between towns along the river. The purpose of the Authority will be to restore fisheries, trapping, and hunting as important and sustainable economic activities. The necessary steps should be to:
— Establish a fisheries reclamation bond issue.
— Buy abandoned industrial sites along river, including dams.

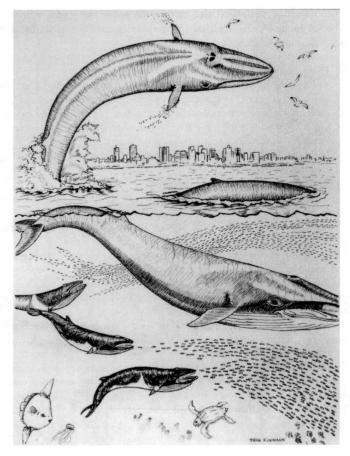

Figure 3.88 Stellwagen Bank was once the habitat for enormous populations of several species of whales. It could be again. (Drawing by Fran Eisemann.)

— Require owners to detoxify and remove reservoir sediments, and then to remove the dams.
— Sell reclaimed sediments and land.
— Eventually sell harvesting rights to commercial fishers and trappers.
• Detoxify sediments in reservoirs and Fowl Meadow using streamside sand beds and treatment plants, starting with existing facility at Foxboro State Hospital.
• Remove dams that impede migration of fish, namely the dams below Fowl Meadow.
• Remove dams that impede normal spring flushing of organic matter, including trees, namely the dams above Fowl Meadow, such as the ones in Canton (Figure 3.89).
• Reclaim wooded banks along river and remove industrial trash.
• Restrain overland flow from storm sewers.
• Estimate value of restored river habitat for:
— Fin fisheries
— Shell fisheries
— Fur-bearing mammals
— Real estate

Figure 3.89 Migratory herring or alewives used to ascend the East Branch of the Neponset River in Canton in enormous numbers. With the removal of the two major dams near the mouth of the Neponset River, and the removal of three dams on the East Branch, the herring could again run as far up as Lake Massapoag in Sharon. (Drawing by Fran Eisemann.)

- Establish sustainable harvesting rates for:
  — Fish
  — Shellfish
  — Fur-bearing mammals
  — Birds
  — Deer
- Restore important species as keystone species, for example:
  — Fish
  — Shellfish
  — Turtles
- Sell harvesting and trapping licenses, and thus provide revenues for operating expenses of programs.
- Establish the Upper Neponset Ecological Facility.

The Upper Neponset Reservoir Ecological Restoration Facility was proposed in 1992 to the town of Foxboro as a major step in restoring the Neponset River to its original health, suitable for fishing and swimming throughout its length. In 1966, the Commonwealth of Massachusetts had established fishing and swimming as the designated uses for the Neponset River. A major impediment to the river's health are the several contaminated reservoirs in the upper valley. Restoration of the river should be started by the construction of the Upper Neponset River Ecological Restoration Facility, which would be used to remove contaminated portions of the reservoir sediments.

These reservoirs contain excessive nutrients and high concentrations of toxic metals that are periodically flushed down the length of the river during storms in the upper basin. The contaminants impede development of normal aquatic life and fisheries all the way to Dorchester Bay. Excessive algae growths in the summer are offensive, damage fisheries and shoreline property values, and make the river and reservoirs unsuitable for swimming.

In 1994, the governor proclaimed a Neponset River Initiative, requesting concerned agencies and citizens to focus on the river. Many activities have begun since then, including several aimed at restoring normal fisheries in Dorchester Bay and the estuary. This includes studies on fish ladders on the lower dams to promote migratory fish populations.

We must not forget that the Neponset River originally supported large populations of smelt, shad, and alewives, as well as trout, bass, and pickerel. There were commercially significant fishing groups in Canton and Norwood before the dams were built. These fisheries were destroyed by industrial dams and pollution that began about 1890, about the same time that Neponset Reservoir was created. It is time they were restored.

3.6.2.3 Low Summer River Flows and Sewage Irrigation in Southeastern Massachusetts

In addition to the natural low flow of rivers in New England during the late summer, two other influences have caused particularly disastrous changes in river ecology during this season. The first is the decrease in the base river flow coming from the aquifer, due to depletion of the aquifer by extraction upstream from wells that supply growing communities with domestic water supplies. Although the effect of this steadily increasing extraction is small when winter and spring snowmelt and heavy rains cause surging floods in the rivers, it becomes very important at the end of the dry summer when there is no rain. Many rivers in New England now go dry by August or September due to upstream extraction.

The second disastrous influence is the overwhelming discharge of treated sewage from large regional sewage facilities, especially when discharging into one of the smaller rivers. In many cases, the sewage flow is several times larger than the low summer flow of the river. Thus, the character of the river depends entirely on the operation of the treatment plant. Even with tertiary treatment devices, the effluent from the treatment plants can be highly unsuitable for aquatic organisms, especially when they are under stress from adverse summer conditions.

Replacing Extracted Flow — To deal with the loss of base flow from rivers due to extraction by municipal wells, there are no easy, complete solutions. The usual solution would be to construct reservoirs to store water for augmentation of the lower summer flows. These reservoirs would have to be at the headwaters of the river system, and would have to capture spring floods for summer release to the dry riverbed. This was the principle behind the design of most industrial dams in New England; thus, those dams that cannot or need not be modified to restore fisheries could instead be operated for augmentation of low summer flows. Neponset Reservoir at the headwaters of the Neponset River is a good example of this, restoring the summer flow to the river that was reduced by the Foxboro town wells near the rim of the aquifer.

There may be other sites in the upper reaches of river systems where summer flow augmentation is needed. These dams should be constructed as part of fisheries restoration programs, and part of the cost of construction and operation should be borne by the municipalities that are causing the depletion of the summer river flow due to extraction from the aquifers.

Sewage Irrigation — The second adverse influence in New England river ecology is the overwhelming sewage flow during the dry summer months, especially where regional facilities are constructed on small rivers. The common response to this second influence in the 1980s, when many of these regional facilities began operation, was to impose extra treatment requirements for the summer. Additional chemical precipitation, sedimentation, filtration, and aeration were added to the normal treatment processes. This was almost sufficient to make sewage into a river, but at considerable operating expense.

There are at least six major sewage plants in southeastern Massachusetts that have to implement extraordinary and expensive processes in the summer because of the low flows in the rivers. These are the Brockton, Milford, Charles River, Rockland, Mansfield, and Attleboro facilities (Table 3.20). The largest of these plants is the Brockton facility, which discharges a mean of 28 cubic feet per second (cfs) into a stream that has a low flow of only 1 cfs. Water quality standards in this stream can only be met theoretically, and the huge summer flow of sewage completely disrupts the normal ecology of the stream and downstream rivers. Discharge of the highly treated sewage to farmland for irrigating crops would be a much more sensible and sustainable use of the sewage, and could provide important revenue from agriculture, as well as reducing the cost of the required treatment.

On rivers where the local geography is suitable, it should be possible to find large tracts of potential farmland that could be gravity irrigated from the outlet of the sewage treatment facility. Gravity irrigation of alfalfa would have a rather high economic return during the dry summers when cattle and horses have only dry pastures to graze. Even if some pumping were required to reach suitable tracts of land, alfalfa commands a good price in certain areas of New England. Thus, if the sewage authority joins with local farmers, the profit on the alfalfa could somewhat reduce the cost of extra treatment required for the small portion of the plant effluent that might still have to go into the river.

Table 3.20   Six Sewage Plants in Southeastern Massachusetts with Extra Requirements for Summertime Operation, Located in Farming Areas Where Sewage Irrigation is Geographically Feasible

Town	Year of construction	Plant discharge, in cfs	Receiving river	Low summer flow, in cfs
Advanced Sewage Treatment Systems				
Brockton	1984	28	Salisbury	1
Milford	1986	7	Charles	1
Rockland	1982	4	French	1
Secondary Sewage Treatment Systems				
Mansfield	1985	5	Three Mile	4
Attleboro	1980	13	Ten Mile	6
Charles River District	1979	7	Charles	6

There are several potential sites that bear investigation in Massachusetts. There are considerable unused state lands downstream of the Mansfield regional sewage facility on the Three Mile River. There is a state farm downstream of the Brockton sewage facility on the Taunton River. There are several state institutions along the Charles River downstream of the regional sewage facilities of Milford, Millis, and Medfield.

Alfalfa can also be fertilized with sludge pellets to feed cows, horses, and managed deer herds. Deer are coming back to the forests of southern New England and should be managed as an additional food source for the region. Sufficient food can be grown for the deer sites along rivers by fertilizing the floodplains with sludge pellets. Perhaps New Englanders can start organizations to "save the deer" by protecting their habitat and ensuring their food sources. A fitting start to the new millennium would be a coordinated campaign across New England to "save the whales, save the herring, and save the deer."

REFERENCES

Allen, Scott, 1992. *The Patriot Ledger.* Harbor Flounder Getting Healthier. May 16–17, p. 1.

Allen, Scott, 1996. *The Boston Globe.* Boston Harbor Cleanup Progresses. April 1, p. 17.

Bacon, George D. and Switzenbaum, Michael S., 1989. Technical Report. Variation of Heavy Metals Concentration in Municipal Sludge and Sludge Compost. Research and Development #87-01-02. Environmental Engineering Report Number 108-89-4. Environmental Engineering Program, Department Of Civil Engineering, University of Massachusetts.

Bickford, Walter E. and Dymon, Ute Janik, 1990. An Atlas of Massachusetts River Systems, Environmental Designs for the Future. Massachusetts Department of Fisheries, Wildlife and Environmental Law Enforcement. Amherst, University of Massachusetts Press.

Bittman, Mark, 1997. *The New York Times.* Harvesting the Vanishing Feast of Winter. February 5, p. B6.

Bley, Patrick W. and Moring, John R., 1988. Freshwater and Ocean Survival of Atlantic Salmon and Steelhead: A Synopsis. Fish and Wildlife Service, U.S. Department of the Interior. Biological Report 88(9).

Bley, Patrick W., 1987. Age, Growth, and Mortality of Juvenile Atlantic Salmon in Streams: A Review. Fish and Wildlife Service, U.S. Department of the Interior. Biological Report 87(4).

Boston Globe, Sunday, 1994. New England Waters. December 11.

Bradsher, Keith, 1996. The New York Times. Cleanup Efforts Mean Fish Are Biting, Again. April 4, p. A-14.

Broad, William J., 1996. The New York Times. Calving of Right Whales Faces New Threats. A Species Fails to Recover. April 9.

Briggs, John C., 1990. Rhode Island Streams: 1978–88, An Update on Water-Quality Conditions. U.S. Department of the Interior, U.S. Geological Survey. Water-Resources Investigations Report 90-4082.

Brems, Lisa, 1994. The Boston Globe. One-Stream-at-a-Time Cleanup Scours the Neponset River. December 25, p. 3.

Cadmus Group Inc., The, 1994. Merrimack River Initiative. Water Connections. Low-Flow Hydrology of the Merrimack River Watershed.

Chesmore, Arthur P., Testaverde, Salvatore A., and Richards, Paul F., 1971. A Study of the Marine Resources of Dorchester Bay. Massachusetts Department of Natural Resources, Division of Marine Fisheries, The Commonwealth of Massachusetts.

Cortese, Anthony D., ScD., 1983. Taunton River Estuary Study. Final Report. Research and Demonstration Project 83-12. Commonwealth of Massachusetts, Department of Environmental Quality Engineering, Division of Water Pollution Control. Second Quarter 1996.

Doeringer, Peter B. and Terkla, David G., 1995. Trouble in Fishing Waters. Bostonia Magazine, Spring 1995.

Dorfman, Richard, 1987. Taunton River Basin Low Flow Profile. Memo of Massachusetts Department of Water and Power Committee.

Druell, Gregory S., Ginsburg, Lisa C., and Shea, Damian, 1991. CSO Effects on Contamination of Boston Harbor Sediments. Massachusetts Water Resources Authority, Environmental Quality Department Technical Report Series Number 91-8.

Dunn, William B., 1994. 1992 Housatonic River Survey. Massachusetts Department of Environmental Protection, Office of Watershed Management. Publication Number 17627-89-50-11/94-1.95-C.R.

Duston, Nina M., Batdorf, Carol A., and Schwartz, Jack P., 1990. Metal Concentrations in Marine Fish and Shellfish from Boston and Salem Harbors, and Coastal Massachusetts. Massachusetts Division of Marine Fisheries, Executive Office of Environmental Affairs.

Elwood, J.R., 1994. Treatment of an Urban Waterway as a Stormwater Mitigation Measure. Northeastern University, Department of Civil Engineering, Boston, MA. Pilot/Demonstration Project Number CP001900-01.

Fierra, David, 1991. Blackstone River Initiative. U.S. Environmental Protection Agency, Massachusetts Division of Water Pollution Control.

Fitzgerald, Brian, 1995. Thar She Blows — Up! Bostonia Magazine, Spring 1995.

Fontaine, Richard A., 1987. Flood of April 1987 in Maine, Massachusetts, and New Hampshire. U.S. Department of the Interior, U.S. Geological Survey. Open File Report 87-460.

Fournier, Paul, 1989. $6B Boston Harbor Cleanup. New England Construction. Boston.

Foxboro Company, 1997, Public Involvement Plan Release Tracking Number 4-11387 Neponset Reservoir, Foxboro, MA. The Foxboro Company.

Franklin, Barbara H., 1992. Status of Fishery Resources off the Northeastern United States for 1992. Conservation and Utilization Division, Northeast Fisheries Science Center, U.S. Department of Commerce. NOAA Technical Memorandum.

Gay, Frederick B. and Frimpter, Michael H., 1984. Distribution of Polychlorinated Biphenyls in the Housatonic River and Adjacent Aquifer, Massachusetts. U.S. Department of the Interior, U.S. Geological Survey. Open File Report 84-588.

Halliwell, David B., 1984. A List of Freshwater Fishes in Massachusetts, Series Number 4. Massachusetts Division of Fisheries Publication Number 13594-14-750-4-84-C.R.

Hanley, Nora E., 1989. The Merrimack River. Part A. Water Quality Data; Part B. Wastewater Discharge Data and Drinking Water Treatment Plant Data; Part C. Water Quality Analysis. Massachusetts Department of Environmental Protection. Publication Number 16, 411-66-25-8-90-CR.

Hogan, Paul M., 1993. Upper Connecticut River 1990 Water Commonwealth of Massachusetts, Executive Office of Environmental Affairs, Department of Environmental Protection, Water Pollution Control Division. Publication Number 17350-28-30-.59 -5/93-C.R.

Horowitz, Ellie, 1985. *Massachusetts Wildlife Magazine,* Special Acid Rain Issue. Volume 35.

Huang, W. and Spaulding, M., 1995. Modeling of CSO-Induced Pollutant Transport in Mt. Hope Bay. American Society of Civil Engineers, *Journal of Environmental Engineering,* 121(7), 1905–1910.

Isaac, R., 1991. POTW Improvements Raise Water Quality. *Water Environment and Technology,* 25, 69–72.

Jobin, W. 1993. Ecology of the Neponset Reservoir in Foxboro at the End of the 20th Century. Rumford River Laboratories, Foxboro, Massachusetts.

Kelley, Barbara G., 1995. The Merrimack Project, Building Partnerships to Prevent Pollution in a Watershed, Summary Report. Commonwealth of Massachusetts, Executive Office of Environmental Affairs. Project Grant X1001588-01-0.

Kennedy, Laurie E., Maietta, Robert J., and Nuzzo, Robert M., 1993. 1992 Housatonic River Tributary Biomonitoring Survey, Assessing Instream Impacts to Biota from Surface Water Supply Withdrawals. Massachusetts Department of Environmental Protection, Division of Water Pollution Control. Publication Number 17331-56-50-1.21-4/93-C.R.

Kennedy, Laurie E., O'Shea Leslie K., Dunn, William J., and LeVangie, Duane, 1994. The Neponset Watershed 1994 Resource Assessment Report. Resource Assessment Project Number 94-1. Department of Environmental Protection, Office of Watershed Management, The Department of Fisheries, Wildlife, and Environmental Law Enforcement Riverways Program.

Kologe, Brian R., 1992. Last of the King of Fishes, The Merrimack River's Lingering Royalty. *Massachusetts Wildlife Magazine,* Spring 1992.

Lambertson, Ronald E., 1994–1995. Connecticut River Basin Anadramous Fisheries Restoration Program. Federal Aid Progress Report. U.S. Fish and Wildlife Service, Connecticut River Coordinator's Office.

Lazell, James D. Jr., 1972. *Reptiles & Amphibians in Massachusetts.* Massachusetts Audubon Society.

Lee, G.F. and Jones, A., 1992, Effects of eutrophication on fisheries, *Lake Line,* 12(4), 13–20.

Massachusetts, the Commonwealth of, 1939. Special Report of the Massachusetts Department of Public Health Relative to the Sanitary Condition of the Certain Rivers in the Commonwealth. House number. 2050. Wright & Plotter Printing Company, Massachusetts.

Massachusetts Marine Fisheries, 1985. Assessment at Mid-decade. Economic, Environmental and Management Problems Facing Massachusetts' Commercial and Recreational Fisheries. Commonwealth of Massachusetts, Division of Marine Fisheries. Department of Fisheries, Wildlife and Environmental Law Enforcement, Executive Office of Environmental Affairs.

Massachusetts Water Resources Authority, 1996. General Review Bonds. Paine Weber, Inc.

MacDonald, Douglas B., 1996. The State of Boston Harbor, The New Treatment Plant Makes its Mark. Massachusetts Water Resources Authority. Technical Report Number 96-6.

McAdow, Ron, 1990. *The Concord Sudbury and Assabet Rivers, A Guide to Canoeing, Wildlife and History.* Bliss Publishing Company, Marlborough, MA.

McKearnan, Sarah, 1990. The Massachusetts Bays Program. Progress 1990. Massachusetts Department of Fisheries, Wildlife and Environmental Law Enforcement.

McMahon, Thomas C., 1969. The Housatonic River. Part C. Massachusetts Water Resources Commission, Division of Water Pollution Control.

McMahon, Thomas C., 1969, 1970. The Charles River. Parts A and B. A Study of Water Pollution. Massachusetts Division of Water Pollution Control. Publication Number 5238.

McMahon, Thomas C., 1970a. Millers River Study. Part A. Massachusetts Water Resources Commission, Division of Water Pollution Control, Commonwealth of Massachusetts. Publication Number 5375.

McMahon, Thomas C., 1970b. Taunton River Basin Industrial Wastewater Discharge Study. Commonwealth of Massachusetts, Massachusetts Water Resources Commission, Division of Water Pollution Control.

McMahon, Thomas C., 1970c. A Pollution Survey of Boston Harbor. Massachusetts Water Resources Commission, Division of Water Pollution Control. Publication Number 5323.

McMahon, Thomas C., 1971a. Report on the Charles River. Part C. Massachusetts Water Resources Commission, Division of Water Pollution Control.

McMahon, Thomas C., 1971b. Report on the Charles River. A Study of Water Pollution. Massachusetts Water Resources Commission, Division of Water Pollution Control.

McMahon, Thomas C., 1972. Assabet River Basin Plan for the Control of Water Pollution. Draft I. Massachusetts Water Resources Commission, Division of Water Pollution Control, Commonwealth of Massachusetts.

McMahon, Thomas C., 1984a. A Comparison of Acute Toxicity Evaluations with Macroinvertebrate Community Analyses. Massachusetts Department of Environmental Quality Engineering, Division of Water Pollution Control, Technical Services Branch. Publication Number 13,668-82-65-7-84-C.R.

McMahon, Thomas C., 1984b. Boston Harbor, 1984 Water Quality and Wastewater Discharge Data. Massachusetts Department of Environmental Quality Engineering, Division of Water Pollution Control. Publication Number #14.270-47-100-12-85-C.R.

McMahon, Thomas C., 1985. Boston Harbor, Water Quality and Wastewater Discharge Data. Massachusetts Department of Environmental Quality Engineering, Division of Water Pollution Control. Publication Number #15905-80-25-4-89-C.R.

Michael, H.J., Boyle, K.J., and Bouchard, R., 1996. Water Quality Affects Property Prices: A Case Study of Selected Maine Lakes. Miscellaneous Report 398, Maine Agricultural and Forest Experiment Station, University of Maine, Augusta, Maine.

Mitchell, John H., 1993. The Renaissance of Small Rivers. Sanctuary. *The Journal of the Massachusetts Audubon Society,* 33, 5–9.

Montgomery, Robert, 1996. Massachusetts' Neponset Lake Concerns Anglers. *Bassmaster Magazine,* May 1996, 29(5), 25–27.

Nealson, Patricia, 1996. *Boston Globe.* Husbanding Alewives. Southeastern Massachusetts Herring Runs Woven into Fabric of Town Life. May 2, p. 27.

O'Leary, C.J., 1987–88. Upper Charles River Surveys. Massachusetts Department of Environmental Protection, Division of Water Pollution Control. Publication Number #15874-76-41-3-89-C.R.

O'Shea, Leslie K., 1989. 1986 Mystic River Survey 1986. Massachusetts Department of Environmental Protection, Division of Water Pollution Control.

Pielke, Roger, 1990. *The Hurricane.* Routledge Books, London.

Quirk, Thomas P., Lawler, John P., and Matusky, Felix E., 1971. System Analysis for Water Pollution Control. Quirk, Lawler & Matusky Engineers. New York Office File 174-1.

Riely, Elizabeth, 1995. Running the River. *Bostonia Magazine,* Spring 1995.

Ries, Kernell G. III, 1990. Estimating Surface-Water Runoff to Narragansett Bay, Rhode Island and Massachusetts. U.S. Department of the Interior, U.S. Geological Survey. Water Resources Investigations Report 89-4164.

Risley, John C., 1994. Estimating the Magnitude and Frequency of Low Flows of Streams in Massachusetts. Water Resources Investigations Report 94-4100. U.S. Department of the Interior, U.S. Geological Survey.

Ross, Robert M., Backman, Thomas W., and Bennett, Randy M., 1993. Evaluation of Habitat Suitability Index Models for Riverine Life Stages of American Shad, with Proposed Models for Premigratory Juveniles. Fish and Wildlife Service, U.S. Department of the Interior. Biological Report 14.

Roth, Charles E., 1978. An Introduction to Massachusetts Mammals. *The 1978 Yearbook of the Massachusetts Audubon Society.*

Rumford River Laboratories, September 1985. A Public Report. Toward a Cleaner Ten Mile River. Foxboro, MA.

Rumford River Laboratories, 1992. Neponset Reservoir Study. Foxboro Conservation Commission and Neponset Reservoir Restoration Committee, Foxboro, MA.

Schevill, William E., 1974. *The Whale Problem, A Status Report.* Harvard University Press. Cambridge, MA.

Simcox, Alison C., 1992. Water Resources of Massachusetts. U.S. Department of the Interior, U.S. Geological Survey. Water Resources Investigations Report 90-4144.

Smith, J. Douglas, 1975. Final Report on the Storrow Lagoon Demonstration Plant. Commonwealth of Massachusetts, Metropolitan District Commission. Process Research Incorporated, Cambridge, MA.

Smith, J. Douglas, 1975. The Impact of the Federal Water Pollution Control Act on the Charles River and Boston Massachusetts, Metropolitan District Commission. Process Research Incorporated, Cambridge, MA.

Smith, J. Douglas, 1975. Second Interim Report on Evaluation of Federal Program for Control of Pollution in the Charles River and Boston Harbor. Commonwealth of Massachusetts, Metropolitan District Commission. Process Research Incorporated, Cambridge, MA.

Smith, J. Douglas, 1974. First Interim Report on Evaluation of Federal Program for Control of Pollution in the Charles River and Boston Harbor. Commonwealth of Massachusetts, Metropolitan District Commission. Process Research Incorporated, Cambridge, MA.

Spehar, Robert L., Anderson, Richard L., and Fiandt, James T., 1978. Toxicity and Bioaccumulations of Cadmium and Lead in Aquatic Invertebrates. U.S. Department of Commerce, Environmental Research Lab. Report Number EPA-600/J-78-016.

Stolzenbach, Keith D., Adams, E. Eric, Ladd, Charles C., and Madsen, Ole S., 1993. Part 1: Transport of Contaminated Sediments In Boston Harbor. Massachusetts Water Resources Authority, Environmental Quality Department Technical Report Series Number 93-12.

Tierney, Susan F., 1993. The Potential for Spread of the Exotic Zebra Mussel (*Driessena Polymorpha*) in Massachusetts. Massachusetts Department of Environmental Protection, Division of Water Pollution Control. Report MS-Q-11.

Tierney, Susan F., 1992. Selected Freshwater Macroinvertebrates Proposed for Special Concern Status in Massachusetts, (*Porifera, Platyhelminthes, Ectoprocta,* and *Mollusca*). Massachusetts Department of Environmental Protection Division, Water Pollution Control. Report Number MS-Q-10.

U.S. Department of the Interior, 1994. Massachusetts Cooperative Fish and Wildlife Research Unit, Annual Report October 1993–September 1994. National Biological Survey University of Massachusetts, Massachusetts Division of Massachusetts Division of Marine Fisheries Wildlife Management Institute.

Wandel, S. William Jr., 1984. Gazetteer of Hydrologic Characteristics of Streams in Massachusetts — Coastal River Basins of the North Shore and Massachusetts Bay. U.S. Department of the Interior, U.S. Geological Survey. Water-Resources Investigations Report 84-4281.

Weatherford, Jack McIver, 1988. *Indian Givers, How the Indians of the North Americas Transformed the World,* Ballantine Books, New York.

Webber, Margo, 1987. Charles River Basin. Massachusetts Department of Environmental Quality Engineering, Division of Water Pollution Control. Publication Number #15,488-70-35-5-88-C.R.

Webber, Margo, 1990. Charles River Bacteria Study, Bacteria Data and Analysis. Massachusetts Department of Environmental Quality Engineering, Division of Water Pollution Control. Publication Number 16,910-129-25-7-91-C.R.

Weld, William F., 1992. Progress 92, Loving Our Bays Through Wisdom, Experience, and Motivation, The Massachusetts Bays Program. Commonwealth of Massachusetts.

Wilbur, C. Keith, 1978. *The New England Indians.* The Globe Pequot Press. Chester, CT.

Zimmerman, Marc J., Grady, Stephen J., Todd Trench, Elaine C., Flanagan, Sarah M., and Nielsen, Martha G., 1996. Water Quality Assessment of the Connecticut, Housatonic, and Thames River Basin Study Unit: Analysis of Available Data on Nutrients, Suspended Sediments, and Pesticides 1972-92. U.S. Department of the Interior, U.S. Geological Survey, Water Resources Investigations Report 95-4203.

CHAPTER 4

The Americas

In this chapter, the conditions of the waters of the rest of the American continent are reviewed, including their histories of dam building, sewage disposal, irrigation, and water management. There is great diversity, and several lessons to be learned, especially from the Wild West and from the greatest metropolis in the world, Mexico City.

4.1 THE NORTH

Although some aspects of its history are similar to those of the U.S., Canada is much less populated, contains vast tracts of wilderness, water is abundant, and thus irrigation is unusual. However, the abundant flow provides many opportunities for hydroelectric power, both along the St. Lawrence River and in the Canadian Rocky Mountains. Perhaps the main hydraulic engineering monument in Canada is the one that connects it with, and divides it from, the U.S.: the St. Lawrence Seaway.

A review is given in this chapter of some of the social and ecological problems of the hydroelectric projects in Canada and the seaway. These projects have generally been engineering and financial successes. Some of their unexpected negative impacts are illustrated in the hope of making future projects even more acceptable to the overall community of the Americas.

The unexpected problems relate to several important aspects of American ecological history. The first is the disregard for indigenous people when "concrete fever" takes over urban dwellers who want to dam rivers in lightly populated areas. The second is the invasion of new territories by aquatic organisms that take advantage of giant canals such as the St. Lawrence Seaway. And finally, an example is given of massive and toxic contamination of the seaway by chemical industries that have been able to avoid their responsibility to remove or detoxify the chemicals they so carelessly dumped.

4.1.1 James Bay of Canada

The southeastern rim of Hudson Bay, including the smaller James Bay at the southern tip, was a primeval wilderness until recently, populated by seminomadic

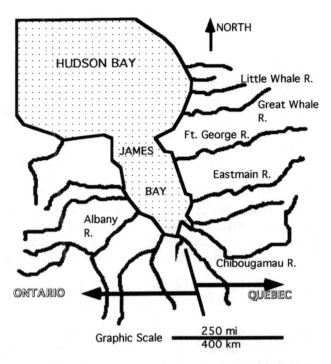

Figure 4.1   James Bay and rivers of the land of the Cree people in central Canada.

Cree people who lived off the fisheries, caribou, and other natural resources (Figure 4.1). But as the population centers of Quebec and the northern U.S. grew, they began to reach northward for resources, including exploitation of the rivers flowing into James Bay and even Hudson Bay for hydroelectric power. The first of these dams was built on La Grande Riviere along the shores of James Bay in northern Quebec.

The James Bay Cree people number about 12,000 at present. They have lived, hunted, fished, and trapped on the eastern shores of James Bay for centuries, in a territory about the size of California. Nine similar groups are joined in the Grand Council of the Crees that govern the Cree Nation. Their resistance to the dam builders during the last decade has been led by the third Grand chief of their nation. In addition to dealing with the negative impacts of the proposed dams in the La Grande Project, their Chief has negotiated directly with Canada over constitutional rights for aboriginal peoples.

The history of dam building around James Bay is remarkable for its initial disregard of the dam builders for the existence and basic needs of the indigenous Cree people of the area, followed by the Cree's strenuous and partly successful rebellion. Some dams were built. But when the Cree people were able to demonstrate the harm being done to their fisheries, their health, and the primeval ecology of the river basins, they were able to obtain at least a temporary reprieve. For the time being, further dam projects have apparently been put on hold.

The initial studies on the first two of the proposed eight dams forming La Grande Project for hydroelectric power along the shores of James Bay, showed a naive disregard for the impact of the dams on migratory fisheries and on downstream fisheries at the saline front coming into Hudson Bay. Protests by the indigenous Cree peoples affected by the proposed dams were ignored, in favor of the power demands of the industrial city people to the south. A more complex issue was also not discovered until the first two dams were operating. An elevated mercury content in the Cree diet was traced to dissolution of mercury from rocks flooded by the reservoirs, then bioaccumulated by the predatory fish at the top of the food chain, the basic element in the Cree diet.

The minimal ecological assessment of the proposed dams was brought to light by further studies on fisheries in the affected rivers. Combined with the health risks, these two projects did considerable damage to the Cree people, and afforded them the data to mount a legal challenge to construction of the additional dams proposed in the project. The successful campaign of the Cree people is one of the few instances in the Americas where indigenous people have been able to resist the "concrete fever" of urban dwellers who demand more and more cheap electricity.

4.1.2 St. Lawrence Seaway

The St. Lawrence Seaway is an engineering marvel and the backbone of heavy industry in North America along the Great Lakes. The ecological history of the seaway, however, is complex and includes some negative impacts, described in the next few pages.

Thunder Bay, Ontario, is Canada's third largest port. It is at Mile Zero of the St. Lawrence Seaway, which stretches from the Great Lakes down the St. Lawrence River to the North Atlantic Ocean (Figure 4.2). The seaway allows ocean-going ships from Chicago on Lake Michigan, and from Duluth and Thunder Bay on Lake Superior, to traverse the former straits and rapids at Sault St. Marie, the Detroit River, Port Colborne, Cornwall, and Montreal, and to finally reach the open ocean at the St. Lawrence Estuary between Nova Scotia and Newfoundland. The maximum characteristics for ships passing the 2345 miles of this system are a length of 730 feet, a draft of 25 foot in fresh water, and a cargo of 25,000 tons.

There is a Hudson Bay Company store in Thunder Bay, located near the point where the Kaministikwia River enters Lake Superior at Fort William. The store is the property of the oldest company in the world. This company has operated continuously since 1670 when King Charles II of England gave it a charter. He did so after a maritime saga that began with incursions into the Americas of Sir Francis Drake and other British pirates looking for silver from South American mines. Three centuries later, the Hudson Bay Company is still the world's largest dealer in furs.

Although founded to establish the fur trade in North America, the port of Thunder Bay is now a channel for the export of wheat from Saskatchewan and Manitoba and lumber from northern forests, using the seaway connection to markets all over the world. Until recently, iron ore and potash were also major exports.

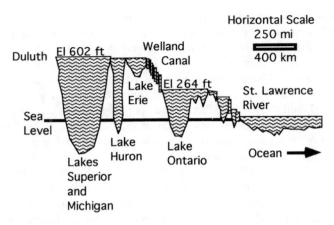

Figure 4.2 Profile of the St. Lawrence Seaway. (Vertical scale is exaggerated.)

4.1.2.1 Lampreys

The purpose of the St. Lawrence Seaway was to allow direct transit of ships from Europe and other continents into the heartland of North America. It accomplished this feat admirably, at least in the warmer months of the year, for the canal is often closed in the frigid winters. However, this transcontinental connection also brought in some unexpected and unwanted passengers. The first major invasion was by the lamprey eel, after World War II. The eels parasitize fish, especially the native lake trout that had been abundant in all the Great Lakes before World War II. The eels attach themselves to adult trout and feed on them until the fish die.

By the 1940s, lake trout populations were declining drastically, nearly disappearing in the 1950s. After research into the life cycles of the lamprey eels in the 1950s, a lamprey control program was initiated in the 1960s that brought the lamprey population down to fairly low levels. Along with intensive stocking and regulation of fishing, control of the lamprey eels allowed the trout and other fish to return to about half their original populations by the 1990s. This was a long-term and very expensive control program, driven by the immense commercial and recreational value of the lake trout.

The control of lamprey eels in the Great Lakes requires application of a toxic chemical to tributary streams around the lakes where the eel larvae develop. This type of chemical control must proceed indefinitely — a costly legacy of the seaway.

4.1.2.2 Zebra Mussels

A second unexpected passenger on the St. Lawrence Seaway was the zebra mussel, which arrived in 1986 in the ballast waters of European freighters. This mussel was first discovered in southern Lake St. Clair near Detroit, at the northwestern tip of Lake Erie (Figure 4.2). Detroit is currently one of the most heavily trafficked ports on the seaway. By 1990, these mussels were spreading eastward through Lake Erie, and fears were expressed that they would spread to the other lakes.

These fears were well founded, and by 1995 the zebra mussel had spread to all of the Great Lakes and to all of the major river systems of central and eastern U.S. With no natural predators, this foreign species occurs in enormous population densities, crowding out native clams and mussels that are already reduced because of habitat stress.

The zebra mussels colonize and clog water intakes for most cities around the Great Lakes, causing enormous costs for their control. Like the lamprey eels, control of these mollusks is through the use of toxic chemicals and will have to occur indefinitely. The real question is how far will they spread in North America? At present, the invaded area continues to grow. In a survey around the entire perimeter of Lake Michigan in 1993, dense populations of adult zebra mussels were discovered at all sampling stations.

4.1.2.3 Love Canal

In order for shipping on the St. Lawrence Seaway to bypass Niagara Falls at the eastern end of Lake Erie, the Welland Canal was dug from Port Colborne on Lake Erie, north to Lake Ontario (Figure 4.2). Niagara Falls was tapped for hydroelectric power, and also preserved as one of the spectacular scenic sites on the Canadian–U.S. border. Despite the Welland Canal bypass, most of the Niagara River continued to flow, providing cheap electricity and also abundant water. This combination of water and power became the basis for a chemical industry, with some unfortunate legacies that were not addressed until recently.

One of the reasons for establishing the federal Superfund program of the EPA for control of hazardous waste sites was the alarming history of a toxic waste dump at Love Canal in the city of Niagara Falls, New York. The partially completed canal around the falls was initiated to provide electrical power to the chemical industries soon after the turn of the century. However, Mr. Love went bankrupt and left his canal unfinished, to be utilized after World War II by Hooker Chemical Company to bury its toxic wastes.

This chemical dump was accompanied by others, including one at nearby Hyde Park, which have leaked dioxin and other toxic chemicals into the Niagara River for generations, causing significant damage to natural aquatic resources such as fish. But Love Canal is most famous because a particularly wet year caused the chemicals to ooze out of the ground, filling the basements of homes constructed over the abandoned site, and contaminating a school belonging to the city. This gruesome event occurred in 1977 and provoked a series of government studies, but no action.

Vigilantes — After 3 years of studies, the desperate local residents kidnapped investigators from the federal EPA and demanded action to protect them and their homes, which they had been forced to abandon.

Based on the stimulus of the citizens' vigilante committee, the federal and state governments then sued Hooker Chemicals, which had changed its name to Occidental

Chemicals. Although the company was willing to assist in the cleanup somewhat, the burden of expense fell on the state and federal governments. The homes were purchased, the school abandoned, and the surrounding area closed down. In 1980 when the EPA established the Superfund program for hazardous waste cleanup, Love Canal was one of the first sites to be studied. By 1982, most of the homes had been purchased by the federal government and the area was thoroughly investigated again by the EPA.

By 1985, the houses were torn down, the canal capped, and drains were placed around the canal to contain the groundwater contamination. A few years later, some improvement was noticed in the groundwater chemistry.

Then, in 1990, the creeks and sewers leading to the Niagara River were cleaned, rehabilitated, and dredged, primarily to remove dioxin contamination. Another EPA study in 1992 determined that the problem was contained, and that people could rebuild in the area again, if they wished.

Industrial Resistance — Throughout this grisly period, from 1977 to 1992, Occidental Chemicals was only willing to assist in part of the remediation, and would not admit liability. Finally, in 1993, they settled with the state of New York; and in 1996, they settled with the federal EPA for payment of less than one third of the damages claimed. A year later, the EPA was still trying to obtain compensation for damages to natural resources such as the fish and other aquatic life that had been contaminated by these toxic chemicals for the last 50 years.

As late as 1995, exotic petrochemicals were still found in fish, crustacea, and marine mammals in the lower St. Lawrence River system, including the estuary. It is certain that many of these chemicals came from the Love Canal area, and from other major chemical industries of the lower Great Lakes. The upper 10 centimeters of sediments in the seaway contained polychlorinated dibenzodioxins and polychlorinated dibenzofurans. These complex and deleterious chemicals were found at concentrations far above normal in plaice. There were irregular patterns of contamination in snow crabs and Nordic shrimp. Stranded beluga whales were found to contain polychlorobiphenyls during 1984 in the St. Lawrence Estuary. The highest concentrations were found in the blubber of these whales, but concentrations of 2000 parts per million (ppm) were found in their milk. Metabolites of DDT were also found in the milk at about the same concentration.

The history of this 19-year mess is not over, but it is clear that the Niagara River and the entire St. Lawrence Seaway to the Atlantic Ocean was subjected to toxic insult for almost half a century. And it was tragically clear that the arcane rules and regulations of the Superfund Act could be easily subverted by the polluters to avoid responsibility and delay compensation.

A cynic would say that the regulations were written by lobbyists for the industries, given the voluntary and slow methodologies prescribed by the Superfund regulations, and the lack of enforcement powers of the regulatory agencies. The regulations certainly were not influenced very much by concern for speedy or thorough remediation of the toxic damage, given that the process will take over 20 years, even for partial remediation.

REFERENCES

Bills, Terry D., Rach, Jeffery J., Marking, Leif L., and Howe, George E., 1992. Effects of the Lampricide 3-Triflouromethyl-4-nitrophenol on the Pink Heelsplitter. U.S. Fish and Wildlife Service, U.S. Department of the Interior. Publication 183, p. 7.

Caron, O. and Roy, D., 1980. The Damming of La Grande Riviere, Quebec. A Critical Period for Aquatic Fauna Downstream of the LG 2 Dam. *Eau du Quebec*, 13(1), 23–24 and 26–28.

Canada Board of Fisheries 1975, Environmental Impact Assessment and Hydroelectric Projects: Hindsight and Foresight in Canada. *Journal of the Fisheries Research Board of Canada*, 32(1), 97–209.

Gilderhus, Philip A., Bills, Terry D., and Johnson, David A., 1992. Method for Detoxifying the Lampricide 3-Triflouromethyl-4-nitrophenol in Streams. U.S. Fish and Wildlife Service, U.S. Department of the Interior. Publication 184, p. 5.

Kuzia, Edward J. and Black, John J., 1985. Investigation of Polycyclic Aromatic Hydrocarbon Discharges to Water in the Vicinity of Buffalo, New York. United States Environmental Protection Agency. EPA-905/4-85-002.

New York Times, The, 1996. Action Urged to Stem Invasion of Species from Ships' Ballast. July 23, p. C-4.

Pope, Phillip E., 1995. Water Quality and the Environment. *The Helm Magazine,* Illinois-Indiana Sea Grant Program, 11(3).

Quiddington, P., 1991. Indians Cheer Halt to Canada's Giant Hydro Scheme. *New Scientist,* 130(1770), 17.

Raphals, P., 1992. The Hidden Cost of Canada's Cheap Power. *New Scientist,* 133(1808), 50–54.

Schultz, Donald P., Harman, Paul D., and Luhning, Charles W., 1979. Uptake, Metabolism, and Elimination of the Lampricide 3-Triflouromethyl-4-nitrophenol by Largemouth Bass (*Micropterus salmoides*). U.S. Department of the Interior, Fish and Wildlife Service.

Shifrin, N., 1997. Gradient Corporation, Cambridge, MA. Personal communication.

4.2 THE WEST

The Wild West of North America, especially the Pacific Northwest and the Rocky Mountains, has been the scene of important recent changes in American ecology. These changes include the destruction of Pacific salmon populations in the Columbia River Basin, the disappearance of the great herds of bison or American buffalo from the Great Plains, and the arrogant displacement of native Americans by massive programs of dam building (Figure 4.3). There have also been a few positive changes, most notably the restoration of elk and deer populations in the Rocky Mountains. Finally, there are some important opportunities as well, including the reuse of sewage for irrigation, the restoration of primeval patterns of seasonal flows by careful operation of hydroelectric dams, and even the possibility of modification or removal of some of the most destructive dams.

4.2.1   Columbia River

One of the clearest and most dramatic devastations of aquatic ecology in the Americas has been the destruction of the salmon and sturgeon fish populations in the

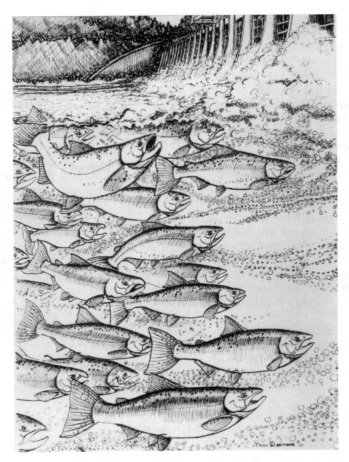

Figure 4.3 Bonneville Dam is the first major impediment faced by Pacific salmon in their annual migration cycle up the Columbia River in the northwestern U.S. (Drawing by Fran Eisemann.)

Columbia River system by uncontrolled dam building, rapacious fishing, unrestricted logging, mining, and other sorts of poorly guided development. It ranks with the destruction of the herds of American buffalo and the global populations of whales as horrible examples of the ecological impact of large industrial and commercial interests.

The Columbia River Basin, including the Snake, Willamette, Salmon, and Clark's Fork Rivers, drains an area of about 400,000 square miles, and has a mean annual flow of about 200,000 cubic feet per second. At present, there are 14 major federal dams on the river, and about 20 others on the tributaries, including the Snake River (Figure 4.4).

When the only human inhabitants of the basin were the original American tribes of the Klamath, Shoshone, and others, chinook salmon from the Pacific coastal waters migrated 1200 miles up the Columbia River every year to spawn around Lake Windemere in Canada, and 600 miles up the Snake River to Shoshone Falls in Idaho,

Figure 4.4 Columbia and Snake Rivers in the Pacific Northwest. Small rectangles denote location of 36 dams in the river system. 1, Bonneville Dam; 2, Chief Joseph Dam; 3, Grand Coulee Dam; and 4, Hells Canyon Dam.

the largest chinook salmon population in the world. These fish were harvested at a sustainable rate for thousands of years before the white immigrants arrived.

Now, after only two centuries of settlement by the immigrants who followed Sacajawea with the Lewis and Clark Expedition, the chinook salmon are listed as "threatened" on the Endangered Species List, with less than 100,000 returning each year at the close of the millennium (Table 4.1). Many of these are unsustainable strains of fish produced in hatcheries. The overall impact on all salmonid fish in the Columbia River during these three centuries has been a reduction to 5% of their original numbers.

Recent reports for the Snake River are particularly horrifying. In 1969 about 100,000 chinook salmon returned to spawn in the Snake River. In 1995 only 8000 returned. Even worse, of the once abundant sockeye salmon population originally inhabiting the Snake River, the number returning in 1996 was 1.

The enormous primeval populations of white and green sturgeon and the Pacific lamprey have also been reduced to 1–3% of their normal numbers. The sturgeon

Table 4.1   Three Centuries of Population History for Aquatic Resources in Columbia and Snake River Basins, Pacific Northwest

Year	Mid-18th Century		Mid-19th Century		Mid-20th Century		End of 20th Century	
	Total	Taken per year	Total	Taken	Total	Taken	Total	Taken
Salmon and steel head trout	380[a]	100[a]		43	10[a]	20	22	Nil
Shad	None		None		1	0.5	4	0.3
Lamprey	Normal				1.5	0.2		0.03
Sturgeon	Normal		6–10[b]	5.5[b]	Nil		0.02	

Note: Total numbers are rough estimates in millions of pounds making run annually. Amount taken is yearly harvest given in millions of pounds.
[a] Assuming average weight of individual = 20 pounds.
[b] Peak harvest of 1892. Followed by collapse of fishery.

populations were nearly extinguished in 1880, and may never recover. Of the migratory fish, only the introduced Atlantic shad have prospered in the river, due to their ability to live in artificial reservoirs (Table 4.1).

The combined effects of the extractive commercial operations, human settlement, and dam building have hurt more than the aquatic resources of the basin, but the impact of the dams on fish has been documented most thoroughly. Dams reduce the migratory fish populations in several ways. The dams and the operating practices for the reservoirs they created:

1. Increase the travel time of adult fish going upstream to spawning grounds, thus reducing spawning success.
2. Increase the travel time of juvenile fish going downstream to the ocean, thus reducing maturation success.
3. Disorient the fish's migratory progress in both directions because of the lack of currents in the reservoirs.
4. Increase salmon losses from predation by other fish in the reservoirs which increases with travel time for the salmon.
5. Raise the average temperature of the water, which causes increased production of predatory fish.
6. Cause death of fish from nitrogen gas bubbles as they pass the dams.
7. Kill fish passing through turbines, valves, and spillways.

In 1990, the Shoshone–Bannock people at the headwaters near Shoshone Falls in Idaho petitioned the U.S. Government to place the salmon on the endangered list because the numbers were too small to even allow for religious ceremonies. By 1991, it was realized that as many as five salmon species may belong on that list.

During the Great Depression of the early 20th Century before most of the dams were built, the migratory fish populations in the river were already in great danger due to over-fishing. The original American fishing tribes had been pushed out long before by commercial enterprises that operated as many as 55 canneries and used all manner of devices to catch the fish, including gillnets, fishwheels, and traps. In one year during the mid-1880s, 43 million pounds of salmonids had been landed.

The canning operations also damaged the salmon habitats because all the fish wastes, as well as excess dead fish that could not be immediately handled by the cannery, were simply dumped into the river, depleting the oxygen and destroying the water quality so necessary to the fish.

Attempts by fishery biologists and dam designers to restore the natural migratory populations have not only been unsuccessful, but in many cases they have caused further damage to the natural populations. Fish ladders are largely ineffective, and hatchery breeding turns out poor survivors that nonetheless compete with the hardy native species for food, habitat, and spawning sites. Crude attempts at transport of the fish around the dams in tankers are largely a waste of money, as one would expect. The salmon populations continue to disappear.

The latest strategy under consideration is to remove some of the dams, especially those on the Snake River, that interfere with the passage of the chinook salmon and others on the Endangered Species List. Less than 50 years after building them, the U.S. Army Corps of Engineers is seriously studying an option that should have been considered in the original designs: decommissioning. In crude terms, the Corps is facing the possibility that some of its magnificent creations may have to be blown up.

The dams under consideration for destruction are Hell's Canyon, Lower Granite, Lower Monumental, and Ice Harbor. These unpleasant prospects could have been avoided if dam engineers had been aware in the beginning that water is a medium for life, and that dams should be designed with a view toward the future.

4.2.2   Colorado River

Although the ecological history of the Wild West has some very sad chapters, including damming of almost all the spectacular rivers and elimination of the magnificent herds of American buffalo, there have also been some promising recent chapters. The most important are the improved management of flows from Glen Canyon Dam on the Colorado River, and the successful management of deer and elk herds in the Rocky Mountains of Colorado. These new approaches in the Wild West offer hope for even crusty New England.

Comprehensive development of the water resources in the Colorado River Basin includes the original Boulder Canyon Project in the lower basin, based on Hoover Dam, which was completed in 1935 about 25 miles southeast of Las Vegas, Nevada (Figure 4.5). In 1956, the Upper Colorado River Storage Project above Lees Ferry, Nevada, was authorized to include Glen Canyon, Flaming Gorge, Navajo, and Curecanti Dams on the Colorado River and tributaries.

When Glen Canyon Dam was constructed on the Colorado River in 1963, the designers wanted to capture and store the torrential spring floods that annually scoured the Colorado River and had, since primeval times, created the Grand Canyon. In the following three decades, the character of the riverbed and the riparian ecology of the river changed, due to the elimination of this annual flood and to the discharge of higher and colder flows while the dam was generating electricity during the normally dry autumn. Prior to the dam, the Colorado River normally diminished in flow and rose significantly in temperature in the autumn.

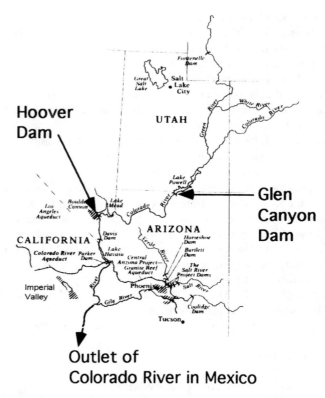

Figure 4.5   Location of Glen Canyon and Hoover Dams on the Colorado River. (Adapted from Reisner, 1986. With permission.)

After the dam was completed, downstream deposits and sandbars soon accumulated in the river, the highest riparian lands and wetlands dried up, and the fish, amphibian, mollusk, bird, and insect population groups shifted to those that were able to colonize steady, cold flow. Those species whose life cycles depended on the variability of seasonal flow in the river, gradually disappeared. The normally warm and muddy river in the summer and early autumn had become a clear, cold stream. The clear water allowed algae to grow profusely on the bottom of the stream, creating a new bottom ecology as well.

4.2.2.1 *Artificial Spring Flood*

In 1996, the Department of the Interior decided to experiment with an artificial spring flood, to determine if they could reestablish some of the original habitat conditions in the Grand Canyon (Figure 4.5). This experiment began with the hope of slowing the loss of endangered native species of fish and other aquatic organisms.

For 7 days in March 1996, an artificial flood of 45,000 cubic feet per second (cfs) was released from the low level outlets in Glen Canyon Dam, causing noticeable rises in the flow immediately downstream. The artificial flood also caused a significant though not drastic change in conditions in the Grand Canyon for several hundred miles down to the next impoundment, Lake Mead, created by Hoover Dam (Figure 4.5). The seasonal levels in Lake Powell and Lake Mead were also changed, due to the artificial release of the additional 16 billion cubic feet of water from Lake Powell during the experimental period.

The artificial flood of 1996 was rather mild in comparison with the normal spring floods occurring prior to construction of Glen Canyon Dam. These natural floods were created by rapid melting of snow, combined with heavy spring rains. In most years, spring floods had reached peak flows of 100,000 to 200,000 cfs. The evaluation of the impact of this artificial flood will take several years, but it is an important step in reestablishing some of the important natural ecology of the Colorado River which has been changed so drastically in the latter part of the 20th Century.

4.2.2.2 Elk and Deer Herds in the Rocky Mountains

Aquatic ecologists should learn from the highly successful programs of the Wild West in the Colorado River Basin for restoration of terrestrial wildlife (Figure 4.6). By combining diverse interest groups, including hunters and range management people, programs along the White River in northwest Colorado have managed to build up enormous and solid populations of elk and deer in the last half-century (Figure 4.5).

The White River elk herd reached peak numbers over 33,000 during the past several decades, with a stable population of 30,000. This allowed a sustainable harvest of about 5000 elk each year (Figure 4.7). The smaller deer coexisted with the elk in the White River Basin, with a stable herd of about 100,000 and an annual harvest of about 10,000 elk.

The key to these successful programs is found in such organizations as the Rocky Mountain Elk Foundation, which purchases crucial habitat areas and protects them from misuse. If the crusty New Englanders tried this strategy, they might be able to save their herring and maybe even the Atlantic salmon.

The mission of the Elk Foundation is to ensure the future of elk, other wildlife, and their habitats. Almost all of the 100,000 members hunt or fish, and understand that both harvesting and conservation are needed to maintain healthy herds. Members work on controlled burning of habitat, brush rejuvenation, and revegetation of winter range land. This nonprofit organization has 475 chapters across the U.S. and Canada. In Colorado alone, the Elk Foundation has raised $3.5 million for conserving and enhancing 150,000 acres of wildlife habitat.

The Elk Foundation focuses on land that protects important winter ranges, calving areas, and migratory corridors. Land donations as well as outright purchases are used to obtain this valuable habitat. One of the newest programs is the acquisition of conservation easements to ensure that elk will have habitats for generations to come. The easements allow private landowners to keep their lands and use them for ranching, while preserving wildlife habitat.

Figure 4.6 Elk along the White River in the Colorado River Basin. (Drawing by Fran Eisemann.)

4.2.3 City of Denver and the South Platte River

The largest city on the eastern slope of the Rocky Mountains has been involved in controversy over shortages of water for its growing population, while ignoring an important opportunity for agricultural reuse of its wastewaters. Because of Denver's location upstream of an immense agricultural area, the reuse of its treated sewage for irrigation should be explored for generating income to help pay for the required treatment. After leaving Denver, the South Platte River runs for hundreds of miles through flat, fertile land, lacking only water to be highly productive (Figure 4.8).

Preliminary studies on the application of Denver's sewage to land has shown that the process introduces fewer contaminants, such as nitrates, to the underlying aquifer than does the practice of discharging secondary sewage to the South Platte River. Furthermore, the careful application of the sewage to crops should remove a large portion of the nitrates, phosphates, and other nutrients. The example in a

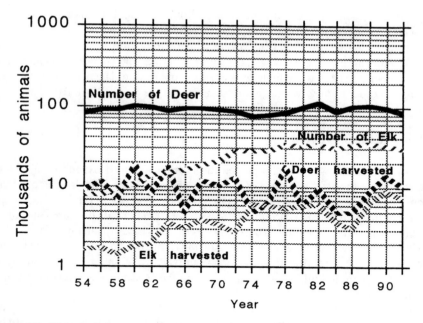

Figure 4.7   History of elk and deer populations in White River Basin of Colorado, 1954–1992.

Figure 4.8   Oahe and Garrison Dams on the Missouri River. These dams are two of over nine major water projects constructed on the mainstem of the Missouri River by the U.S. government. (Adapted from Reisner, 1986. With permission.)

following chapter on the reuse of sewage from Mexico City shows that sewage irrigation is not only possible, it is highly profitable.

Eastern Colorado suffers from severe winters, thus limiting the reuse of sewage to less than half the year. However, the summer period when the sewage can be used is also the dry season, when high degrees of treatment would be needed for Denver's sewage in order to meet water quality standards in the South Platte River. Thus, the savings in treatment costs could be considerable.

4.2.4 Missouri River

Sacajawea, the woman who guided the Lewis and Clark Expedition across the Northwest to the Pacific Ocean, came from the Shoshone–Mandan people living along the upper Missouri River. Generations later, her few surviving descendants were confined to the Fort Berthold Reservation, along with the remnants of the Arikara and Hidatsa peoples. Those who were not killed by smallpox or the U.S. cavalry were able to make a decent living on the fertile floodplain that passed through their reservation. Unfortunately, they stood in the way of an ambitious plan by the U.S. government to build another large dam on the Missouri River just upstream of Bismark, North Dakota.

In a joint effort by the U.S. Corps of Engineers and the U.S. Bureau of Reclamation, nine major dams were built on the Missouri River upstream of Kansas City. Although these are federal agencies, it appears there is no guarantee they obey federal law, or even follow common decency. But their disregard for the needs and rights of the original Americans includes an incredible episode from 1948 that, unfortunately, shows that the U.S. government can be as callous toward indigenous peoples as was the Canadian government toward the Cree people. And the following chapters on South America will show that this is not an isolated national quirk, but apparently a universal tendency of dam builders.

4.2.4.1 Garrison Dam and the Three Tribes

Two of the last dams built on the Missouri River were the Oahe and Garrison Dams (Figure 4.8). Oahe Dam was built by the U.S. Corps of Engineers and is the largest earthfill dam in North America. It began generating electricity in 1962, slightly upstream of Pierre, South Dakota.

Although Oahe Dam is larger, Garrison Dam is 200 feet high and 2 miles long, completed in 1954 slightly upstream of Bismark, North Dakota. Garrison Dam was supposedly designed for hydroelectric power, flood control, and irrigation, but the third function has never been achieved, and siltation from the muddy Missouri River has decreased its storage capacity considerably. Given the disastrous floods of 1995 at the junction of the Missouri and Mississippi Rivers, it is not clear how much value the dam had for flood control either. It is unfortunate that there were many indications that these large dam projects were often the result of dam builders' ambition, and not based on rational analysis.

Although the placement of dams on the Missouri River was carefully designed to avoid flooding of any cities of the immigrant American population, no such consideration was given to the original Americans, the three tribes of the Mandans, Arikara, and Hidatsa. The urban areas of Bismark, Pierre, Chamberlain, and Williston were all carefully protected by placement and spillway adjustments of Oahe, Case, Big Bend, and Fort Randall Dams on the Missouri River. But Garrison Dam, on which the spillway had been lowered 20 feet in order not to flood Williston, North Dakota, would flood the entire reservation of the three tribes, including some of the best winter range land in North Dakota. The three tribes were forced to leave the land they had occupied for many hundreds of years before the white men came.

A Brutal Resettlement Plan — The provisions for their resettlement were hard to believe, even by people who were accustomed to unkept treaties and promises. Besides being displaced to higher land that grew little grass for their cattle, they would not be permitted to fish in the reservoir, nor could their cattle drink from it or graze near it. They were not even permitted to remove the trees that would be flooded and killed by the rising waters of the reservoir. Many of the restrictions were clearly revenge for a public protest the Mandans made when they first learned their land would be taken.

Their complaints went unheeded. The business council of the three tribes pointed out that over 1500 of their people would be displaced. They were all stockmen and their living depended on production of cattle, yet the lands to be flooded were practically all the lands they had for feed production or winter pasture.

But the dam was built and the three tribes have never recovered, being displaced to arid land far from the river and forced to live in the small white towns of North Dakota. At the signing of the bill in Washington, D.C. disposing of the three tribes, the leader of the tribal business council wept openly, foreseeing the end of his people.

But to me, the final insult came from the U.S. officials who gave the lake formed by Garrison Dam the name of the first Mandan befriended by white explorers. They called it Lake Sacajawea.

REFERENCES

Anadramous Fish, 1994. Appendix C-1. Columbia River System Operation Review. Draft Environmental Impact Statement. Bonneville Power Administration. U.S. Army Corps of Engineers. U.S. Department of the Interior. Bureau of Reclamation.

Butler, David L., 1991. Reconnaissance Investigation of Water Quality, Bottom Sediment, and Biota Associated with Irrigation Drainage in the Gunnison and Uncompahgre River Basins at Sweitzer Lake, West-Central Colorado, 1988–89. U.S. Department of the Interior, U.S. Geological Survey. Water-Resources Investigations Report 91-4103.

Colorado Department of Public Health and Environment, 1994. Status of Water Quality in Colorado. State of Colorado, Denver.

Egan, Timothy, 1991. *The New York Times*. Fight to Save Salmon Starts Fight over Water. April 1, p. A-1.

Frazier, Deborah, 1996. *Colorado & The West*. Colorado Water Worries. June 23, p. 21A.

Gaggiani, Neville G. 1991. Effects of Land Disposal of Municipal Sewage Sludge on Soil, Streambed Sediment, and Ground- and Surface-Water Quality at a Site Near Denver, Colorado. U.S. Department of the Interior, U.S. Geological Survey. Water Resources Investigations Report 90-4016.

Hartman, Todd, 1996. *The Denver Post.* Cleanup Standards for River Stir Concern, Pristine Alamosa? December 15, p. 3C.

Jackson, D., 1988, *Great American Bridges and Dams.* The Preservation Press, Washington, D.C.

Karp, Catherine A. and Tyus, Harold M., 1989. Habitat Use and Streamflow Needs of Rare and Endangered Fishes, Yampa River, Colorado. U.S. Fish and Wildlife Service, U.S. Department of the Interior. Biological Report 89(14).

Martineau, Laura, 1996. *Gunnison Country Times.* Dominguez Project Would Flood Lower Gunnison River. July 25, p. 5.

Meeker Herald, 1996. How Big Were the Deer Herds 20 Years Ago? September 26, p. 8.

Miller, Dave, 1996. *Valley Chronicle.* Putting Our Waters in Dire Straits. February 13, p. 4.

New York Times, National, 1996. Artificial Flood Created to Rejuvenate The Grand Canyon. March 27, p. 5.

Pescod, M.B., 1992. Wastewater Treatment and Use in Agriculture. Food and Agriculture Organization of the United Nations. FAO Irrigation and Drainage Paper 47. Rome.

Patricia, A., Skaar, Donald R., and Knight, Denise E., 1987. Factors Affecting the Mobilization, Transport, and Bioavailabilty of Mercury in Reservoirs of the Upper Missouri River Basin. U.S. Department of the Interior, Fish and Wildlife Service. Fish and Wildlife Technical Report 10.

Raleigh, Robert F., Miller, William J., and Nelson, Patrick C., 1986. Habitat Suitability Index Models and Instream Flow Suitability Curves: Chinook Salmon. Biological Report 82(10.122). National Ecology Center, Division of Wildlife and Contaminant Research, Fish and Wildlife Research, U.S. Department of the Interior.

Reisner, Mark, 1986. *Cadillac Desert, The American West and its Disappearing Water.* Penguin Books, New York.

Rennicke, Jeff, 1985. *The Rivers Of Colorado. 1. The Colorado Geographic Series.* Falcon Press Publishing Company, Billings, Helena, Montana.

Stevens, L. and Wegner, D., 1995. Changes on the Colorado River: operating Glen Canyon Dam for Environmental Criteria. U.S. Department of the Interior, Bureau of Reclamation.

Stevens, Williams K., 1994. *The New York Times.* Dwindling Salmon Spur West to Save Rivers. November 15, p. C-1.

Swanson, D., 1997. *The Denver Post,* Revamped River Unkind to Fish, March 12, p. 13A.

Tobin, Robert L. and Hollowed, Caroline P., 1990. Water Quality and Sediment-Transport Characteristics in Kennedy Reservoir, White River Basin, Northwestern Colorado. U.S. Department of the Interior, U.S. Geological Survey. Water Resources Investigations Report 90-4017.

Trembly, Terrence L., 1987. Opportunities to Protect Instream Flows in Colorado and Wyoming. U.S. Department of the Interior, Fish and Wildlife Service. Biological Report 87(10).

U.S. Army Corps of Engineers 1994. Wildlife Appendix N. Colombia River System Operation Review. Draft Environmental Impact Statement. Bonneville Power Administration. U.S. Army Corps of Engineers. U.S. Department of the Interior. Bureau of Reclamation.

U.S. Department of the Interior. 1971–86. Aquatic Ecology Studies of Twin Lakes, Colorado 1971–86, Effects of a Pumped-Storage Hydroelectric Project on a Pair of Montane Lakes. U.S. Department of the Interior, Bureau of Reclamation. A Water Resources Technical Publication, Engineering and Science Monograph Number 43.

Waddell, K.M., Freethey, G.W., Susong, D.D., and Pyper, G.E., 1991. Review of Water Demand and Utilization Studies for the Provo River Drainage Basin, and Review of a Study of the Effects of the Proposed Jordanelle Reservoir on Seepage to Underground Mines, Bonneville Unit of the Central Utah Survey, U.S. Department of the Interior. Open File Report 91-514.

Wydoski, Richard S., Gilbert, Kim, Seethaler, Karl, McAdam, Charles W., and Wydoski, Joy A., 1980. Annotated Bibliography for Aquatic Resource Management of the Upper Colorado River Ecosystem. U.S. Department of the Interior, Fish and Wildlife Service. Resource Publication 135.

Yampa Valley Newspaper, 1997. Elk Foundation Works to Conserve the Yampa Valley. March 22, p. 1.

4.3 THE CARIBBEAN REGION

Tropical ecology involves faster rates of biochemical reactions and biological processes than does the ecology of temperate climates; thus, there are some unique characteristics of aquatic ecology in the tropics, when compared to North America. Examples are given in this section of reservoir ecology on the island of Puerto Rico in the eastern Caribbean Sea, and of sewage reuse in the high semitropical valley near Mexico City. While the ecology is different from that of North America, the Caribbean and Mexican examples are useful in illustrating many of the same problems and solutions as those in previous chapters.

4.3.1 Puerto Rico

Although it is a tropical island populated by people of mixed American, African, and European origins, and typifies the ecology of the Caribbean Sea and Central America, Puerto Rico is nonetheless a possession of the U.S. and is integrated into the U.S. economy and political system. When the U.S. Army took Puerto Rico from Spain at the turn of the century, the U.S. soldiers were soon followed by U.S. investors, their agronomists, and their engineers, intent on using the tropical ecology and cheap labor to produce cash crops such as sugarcane.

The agricultural development of the dry coastal plains of Puerto Rico inevitably resulted in construction of dams on the south and northwest coasts where rainfall often fails. The dams were placed in the mountains to take advantage of more abundant rainfall, and systems of canals carried the water through various regulating reservoirs to the main irrigation canals on the flat coastal plains (Figure 4.9). Unfortunately, for the future of Puerto Rico, many of the dams diverted water away from the population centers in the north, and instead brought it south to the dry coastal plains.

In later years, especially after World War II, several additional dams were constructed even higher in the mountains, and hydroelectric energy was produced as the water fell to the coastal plains, especially on the western end of the island. These dams supplied the base of the island's electrical power system in the early days of this century. About mid-century, the demand for electricity outgrew the supply from these dams, partly due to industrial needs. However, another major

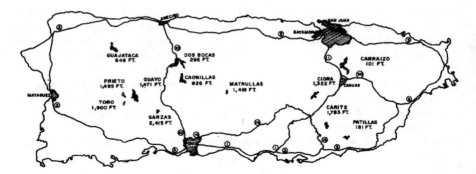

Figure 4.9   Major dams and reservoirs on the island of Puerto Rico in the eastern Caribbean
Sea. The elevation of the spillways is given in feet above sea level. Major highway
numbers are also indicated between principal cities.

source of the rising demand was the rapid population increase, especially on the
north coast around the principal harbor and capital city, San Juan.

By the 1980s, there were 30 of these small dams, used primarily for irrigation,
hydroelectric power, and domestic water supply (Figure 4.9). The oldest was Carite
Dam, constructed on the very headwaters of the La Plata River in 1913 for irrigation
and power; and the newest was La Plata Dam in Toa Alta, constructed to supplement
the supply of domestic water for San Juan in 1973 (Figure 4.10). Both dams were
relatively small, storing only 10,000 to 30,000 acre-feet of water (Table 4.2).

An important aspect of this water resource development in Puerto Rico is the
change in water needs during the 20th Century, resulting in dams and reservoirs that
no longer serve the original intent of their designers. Most dams, reservoirs, and
canal systems in Puerto Rico were designed for 50-year life spans, yet some of them
are now over 80 years old. The large majority have passed their allotted 50-year
age. That they still stand is a tribute to "large safety factors and the grace of God,"
a euphemism to cover the uncertainties of design information at mid-century when
these dams were built. But sugarcane is no longer an important crop in Puerto Rico,
irrigated agriculture is largely gone, and thus the flow in the canals has ceased. The
hydroelectric systems installed after World War II are too small, too far away from
the load centers, antiquated, or even rusted shut. The reservoirs still trap water and
sediments, although a few have so much sediment they no longer trap any water,
such as the Coamo and Comerio Reservoirs (Table 4.2).

Water in modern Puerto Rico is needed primarily for drinking and domestic uses.
Agriculture is gone and electricity is supplied by oil-fired steam generators. But the
exploding population now exceeds 4 million people, and droughts have frequently
provoked water rationing in the past decade. The few reservoirs near San Juan are
eutrophic, filling with sediment and requiring continuous dredging. At times, the
water supply system is in a chaotic state. The most severe periods of water rationing
in San Juan were recently, in the dry years of 1995 and 1996.

This chaos, with multiple dams in the mountains full of water and huge cities
along the coast with great thirsts, was certainly never imagined by the dam designers
of 1913, who proudly watched the concrete being poured and imagined the unlimited

Figure 4.10   La Plata Dam in Toa Alta. The reservoir was already eutrophic in 1975 due to treated sewage coming from the Town of Comerio upstream. (Photo by Guillermo Sosa.)

future use of their creations for irrigation and power (Figure 4.11). The complex of Yahuecas, Guayo, Prieto, Toro, Luchetti, and Loco Dams was constructed in the central mountains to supply water to the southwest corner of the island in the 1950s for power production and irrigation in the Lajas Valley. The water and power is now needed on the north coast of the island where all the people live, but the Lajas system takes the water in the wrong direction.

The reason the dam engineers did not imagine this chaos at the turn of the century is because they were mistakenly committed to the 50-year design life as a rational element in planning of dams. They probably did not expect to be alive in 50 years, so why should they worry about their dams after that? Were any of them aware that the pyramids in Egypt were built to function for over 5000 years?

The Lajas Valley system was small but included the largest hydroelectric generating station in Puerto Rico, below Guayo Dam (Figure 4.11). Power plant #1 had a capacity of 25 megawatts, utilizing the large pressures and storage from the interconnected mountain reservoirs (Table 4.3). Below Luchetti Dam, power plant #2 had a capacity of 16 megawatts.

Consider the magnitude of the electricity produced by all the hydroelectric dams in Puerto Rico in comparison with the current rate of use. All the dams combined could produce about 100 megawatts in 1976. If they are still functioning at the close

Table 4.2   Dams and Reservoirs in Puerto Rico, 1976

Name of dam or reservoir, purpose	Storage volume, in acre-feet	Municipality	Year constructed
Adjuntas, power	465	Adjuntas	1950
Caonillas, power	49,000	Utuado	1948
Carite, domestic supply and irrigation	11,300	Guayama	1913
Carraizo, domestic supply and power	20,000	Trujillo Alto	1954
Cartagena Lagoon, wildlife	70	Lajas	Natural
Cidra, domestic supply	5,220	Cidra	1946
Coamo, irrigation	Abandoned	Santa Isabel	1914
Comerio 1, domestic supply	Abandoned	Comerio	1913
Comerio 2, domestic supply	Abandoned	Comerio	1913
Dos Bocas, power	32,000	Arecibo	1942
Garzas, power and irrigation	4,700	Adjuntas	1943
Guajataca, irrigation	32,600	San Sebastian	1929
Guayabal, irrigation	10,000	Guayabal	1913
Guayo, power and irrigation	17,400	Lares	1956
Guineo, power and irrigation	1,860	Ciales and Orocovis	1931
Jordan, power	Diversion	Utuado	1950
La Plata, domestic supply		Toa Alta	1973
Las Curias, domestic supply	1,100	Rio Piedras	1946
Loco, power and irrigation	1,950	Yauco	1951
Luchetti, power, irrigation	16,500	Yauco	1952
Matrullas, power, irrigation	3,000	Orocovis	1934
Patillas, irrigation	14,500	Patillas	1914
Pellejas, power	152	Adjuntas	1950
Prieto, power and irrigation	700	Lares	1955
Rio Blanco, power		Naguabo	
Toa Vaca, domestic supply and irrigation	33,124	Villalba	1972
Toro, power and irrigation	100	Maricao	1955
Tortugero Lagoon, wildlife		Manati	Natural
Vivi, power	277	Utuado	1950
Yahuecas, power	1,800	Adjuntas	1956

of the century, they probably would supply less than 1% of the total demand for the island, which has a population around 4 million. Events have overtaken the dams and canals, making them irrelevant. Worse, they divert precious water to the wrong parts of the island.

4.3.1.1   Dams, Reservoirs, and Water Quality

Besides shortsighted design aspects, many of the dams and reservoirs are now experiencing water quality problems that make them almost unusable for domestic water supply, the major water need in Puerto Rico. A rapid survey of basic water quality in the 30 reservoirs in 1976 showed that the most important reservoir supplying the largest city on the island had some of the worst water quality. The waters of Lake Carraizo, which are used for San Juan domestic supply, had a color and turbidity of 22 and 30 standard units, respectively, and total phosphates of 0.18 parts per million (ppm) (Table 4.4). Over-fertilization occurs when the total phosphate

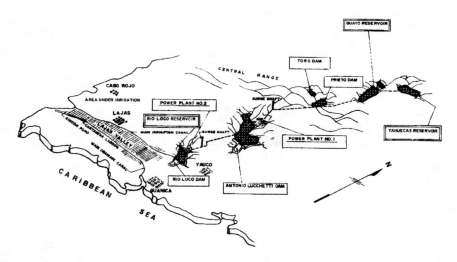

Figure 4.11 Lajas Valley irrigation system. This complex of six mountain dams supplies hydroelectric power and irrigation water to the southwest corner of Puerto Rico.

Table 4.3 **Hydroelectric Reservoirs in Puerto Rico Ranked by Power Production Capacity, 1976**

Reservoir	Power production, in kilowatts	Storage volume, in million cubic meters
Yahuecas, Guayo, Prieto, and Toro diversion, supplying Lajas power plant #1	25,000	18
Dos Bocas	18,000	39
Jordan, Adjuntas, Pellejas, and Vivi diversions to Caonillas	17,600	61
Luchetti, supplying Lajas power plant #2	16,000	15
Garzas	12,240	5.8
Matrullas	8,640	3.6
Carraizo	2,600	25
Guineo	1,920	2.2
Totals	102,000	170

concentrations exceed 0.05 to 0.10 ppm. Acceptable values for color and turbidity in drinking water are usually less than 5 standard units.

The values of the water quality parameters in the lakes of Puerto Rico do not indicate the water quality one would expect from a source for drinking water. Two other reservoirs that showed evidence of severe eutrophication were Adjuntas and Comerio 1 reservoirs, also immediately downstream of large towns. The total phosphates in Adjuntas and Comerio 1 reservoirs were 0.25 and 0.27 ppm, respectively (Figure 4.12).

Additional 1-day surveys in four of the major lakes in 1976 showed marked differences in their degrees of eutrophication and the amount of dissolved oxygen. On the day of these surveys in 1976, Lake Carraizo had only 3 ppm of dissolved

Table 4.4   Water Quality of Reservoirs in Puerto Rico, 1976

Lake	Number of samples	Color, in standard units	Turbidity, in standard units	Total phosphates, in ppm as P
Adjuntas	3	6.7	1.1	0.25[a]
Caonillas	21	8.9	19.9	0.04
Carite	5	7.8	8.0	0.04
Carraizo	6	21.7	30.5	0.18[a]
Cartagena	Not done	Coastal lagoon and swamp		
Cidra	6	8.6	2.8	0.02
Coamo	Not done	Filled with silt, abandoned		
Comerio 1	2	10.0	6.5	0.27[a]
Comerio 2	2	10.0	10.5	0.09
Dos Bocas	6	11.3	6.9	0.08
Garzas	16	7.3	2.2	0.02
Guajataca	2	6.5	1.4	0.01
Guayabal	6	11.7	5.2	0.06
Guayo	10	10.0	2.7	0.02
Guineo	6	5.0	1.6	Not done
Jordan	2	6.0	12.0	0.01
La Plata	2	9.0	9.2	0.04
Las Curias	2	9.0	6.8	0.01
Loco	6	5.3	5.3	0.04
Luchetti	23	5.3	3.0	0.04
Matrullas	8	5.0	2.2	Not done
Patillas	2	12.5	10.1	0.02
Pellejas	2	6.5	15.3	0.01
Prieto	1	12	0.4	0.01
Rio Blanco	Not done	Very small diversion dam		
Toa Vaca	13	6.1	1.8	0.09
Toro	1	5	10.6	Not done
Tortugero	Not done	Brackish coastal lagoon		
Vivi	3	6.3	1.9	0.01
Yahuecas	4	10.5	75.0	0.04

[a] Evidence of severe over-fertilization.

oxygen in the upper 5 meters, while Lake Guajataca had a super-saturation of oxygen, exceeding 11 ppm (Figure 4.13). Although it was not evident in Lake Carraizo, the other four lakes had a clearly well-mixed surface layer of about 4.5 meters depth, in which the wind maintained the dissolved oxygen and other material at a fairly uniform concentration. Lakes Dos Bocas and Patillas had dissolved oxygen concentrations in the surface layers of about 7 ppm, the normal saturation concentration (Figures 4.13 and 4.14).

Extensive additional water quality surveys, conducted quarterly over a 3-year period, gave strong indications that Lake Carraizo was contaminated with sewage, explaining most of its differences in water quality with the other lakes (Figure 4.15). Comparing Lake Carraizo with Lake Guajataca (one of the cleaner lakes), it was evident that phosphate nutrients, turbidity, and color were all very high in Carraizo. In Carraizo, they were 0.28 ppm, 18, and 23, respectively, while the corresponding parameters in Guajataca were 0.01 ppm, 7.6, and 8.4 (Table 4.5).

Figure 4.12 The small Comerio 1 reservoir on the La Plata River was abandoned long ago as a public water supply source due to sewage contamination from the town of Comerio. (Photo by Guillermo Sosa.)

As one would expect, the high color and turbidity in Carraizo reduced the depth of sunlight penetration or the depth of the photic zone to 0.5 meters as read by the Secchi disk, while the low color and turbidity in Guajataca corresponded to a deeper photic zone of 2.2 meters. In this photic zone, photosynthesis and respiration in Lake Carraizo were proceeding at 2.4 and 2.0 ppm of dissolved oxygen per day, respectively, a much higher rate than the 0.6 and 0.5 ppm observed in the photic zone of Guajataca (Table 4.5).

The low dissolved oxygen, high color, turbidity and photosynthesis, and the high counts of coliform bacteria all characterized the water in Lake Carraizo as treated sewage, rather than clean surface runoff from rainfall. This was likely due to the large number of municipal sewage plants in the basin draining to the Loiza River and the Carraizo Reservoir, including that of the large city of Caguas in central Puerto Rico (Figure 4.9). It is unfortunate that a principal water supply for San Juan is also the recipient of so much degraded sewage. It is even more unfortunate that the reservoir probably also contains the industrial wastes from several pharmaceutical industries also located in the Loiza watershed.

The water quality of Lake Carraizo clearly indicated the need for source reduction in industrial and municipal sewage systems upstream, and the increasing difficulty in supplying clean water to San Juan and the metropolitan area.

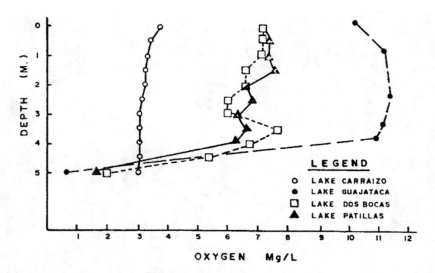

Figure 4.13 Vertical profiles of dissolved oxygen in four major lakes of Puerto Rico, 1978. Concentrations of dissolved oxygen are given in milligrams per liter, as a function of depth below the water surface in meters.

Figure 4.14 Dos Bocas Dam on the Arecibo River. The reservoir often has problems with floating vegetation blocking the spillway, due to fertilization from upstream sewage of the town of Utuado.

Figure 4.15   Lake Carraizo Dam in Trujillo Alto on the Loiza River. This dam and reservoir were originally the source of the entire San Juan water supply. Severe eutrophication problems, including a mat of floating vegetation, have resulted in recent years from sewage contamination from the city of Caguas upstream. (Photo by Guillermo Sosa.)

4.3.1.2   Conclusions for the Island

Because the island of Puerto Rico is surrounded by the salty Caribbean Sea, it has no alternative sources from which to obtain the fresh water needed for domestic purposes, except from the rain that falls on the island. Thus, the water supply is finite. Because human population growth has been exponential in Puerto Rico, the clash of supply and demand for water is strikingly apparent in the water shortages of the past decade. These shortages will not go away; they will get worse. For this reason, management of water resources may become the crucial item in Puerto Rico's ecology. History cannot be turned back in Puerto Rico, but the mistakes now becoming so painfully evident there should be used by planners and water engineers on other islands and in other countries of the Americas, to avoid repetition.

Given the immense importance of water in Puerto Rico's future, the progressive decline and near abandonment of the 30 dams and reservoirs on the island clearly portray the need for a change in principles of dam and reservoir design. Dams and reservoirs must be part of a sustainable system of water management, not single-purpose monuments expected to stand alone for their design life and then suddenly

Table 4.5   Mean Water Quality Over 3 Years of Quarterly Sampling on Four Major Reservoirs in Puerto Rico, 1975–1978

Water quality parameter and units	Lake Carite	Lake Carraizo, showing severe contamination	Lake Guajataca	Lake Patillas
Chlorides, ppm	7.7	16.6	6.6	9.6
Hardness, ppm as $MgSO_4$	29	126	224	29
Total phosphates, ppm as P	0.01	0.28	0.01	0.01
Nitrates and nitrites, ppm as N	0.03	0.33	0.47	0.14
Iron, ppm	0.16	1.02	0.11	0.07
Turbidity, su	3.2	18.3	7.6	9.6
Color, su	9.8	23.0	8.4	10.6
pH	7.18	7.24	7.36	7.72
Dissolved oxygen, ppm	7.0	5.0	7.6	7.4
Coliform bacteria/100 ml	155	17,604	102	399
Secchi disk reading meters	1.6	0.5	2.2	1.8
Temperature °C	23	26	26	29
Productivity, ppm O_2/day = P	0.7	2.4	0.6	0.9
Respiration, ppm O_2/day = R	0.5	2.0	0.5	0.8
Ratio of P/R	1.4	1.2	1.2	1.2
Total phytoplankton count/milliliter	17.5	18.2	Not done	29.4
Chlorophyll-a, ppm	Not done	12.9	Not done	Not done

vaporize. They must have a flexible, sustainable design, or they must have an exit or decommissioning strategy. And the inevitable clash between a fixed supply of water against a rapidly rising demand must be faced from the first day of conceptual design.

Many of the decisions required to reconcile these opposing trends must be made by the community at large, such as limits on populations. But many decisions must also be made by the engineers and planners who design water projects, especially dams and sewage plants.

Water supply reservoirs placed immediately downstream of urban centers must be designed with low-level flushing ports for passage of sewage sediments. The sewage from upstream towns must eventually bypass the reservoir and be discharged to rapidly flowing reaches downstream, or else be recycled. This will entail a large loss in flow for the reservoir, and must eventually be compensated by water conservation or construction of additional dams and storage elsewhere.

Industrial, agricultural, and certain domestic uses of water will have to be satisfied with highly treated domestic sewage to minimize the increasing demand for pure water for human consumption, which must take priority.

Hydroelectric dams and their reservoirs constructed in the mountains far from cities and towns must also be located or interconnected so that eventually their storage capacity can be used for drinking water when the turbines are abandoned. Projections of electrical power needs and water supply needs must be carried beyond 25 to 50 years, and must include the exponential rise of human populations and their water consumption requirements.

Reservoirs for domestic water supply must be designed with the recognition that growth of human populations in the watershed will cause increasing sedimentation and eutrophication of the reservoir, gradually reducing its capacity and the quality

Figure 4.16 Lake Cidra in central Puerto Rico. Despite its safe location, this lake receives a great deal of contaminants from agricultural activities along its shores, which must be controlled. (Photo by Guilllermo Sosa.)

of the water it stores. Human activities such as road building and agriculture, as well as subsurface disposal of sewage, inevitably raise the nutrient level in the groundwater, and eventually the rivers and lakes. This increase in nutrients hastens the naturally slow process of eutrophication.

Thus, the location of such dams must be coordinated with the location of human housing and urban development in order to protect the water stored in the reservoir. The reservoirs should generally be far upstream of the choice places for housing. One of the few water supply reservoirs optimally located to avoid urban contamination is Lake Cidra in the central mountains (Figure 4.16). Hopefully, it will continue to supply safe water to the town of Cidra for years to come.

4.3.2 Mexico

Port cities such as New York, Boston, and San Francisco have followed cyclical patterns of unsustainable responses to contamination of their coastal waters based on large construction projects that gathered all wastes for central treatment and discharge to surface waters. Compare this with Mexico City, which disposed of its sewage by crop and tree irrigation, thus avoiding the cost of treatment and producing crops of continually increasing yield and value.

Mexico City is one of the largest cities in the world, with an estimated population of 18 million people by the end of the millennium. It is in a high, semidesert area with an elevation of 2000 meters above sea level and about 400 millimeters of rain annually (in the summer). The sparse rain falls only in July and September; thus, irrigation is essential for crop production. Given sufficient water, the subtropical climate allows the cultivation of some crops throughout the year, with as many as nine alfalfa harvests possible. This favorable situation was probably the impetus for development of the first informal system of sewage irrigation in the region.

Irrigation of downstream farmlands began in the upper Tula River basin in 1886, utilizing sewage and stormwater that flowed by gravity from Mexico City. The Tula River is a tributary to the larger Panuco River, which discharges to the Gulf of Mexico at Tampico (Figure 4.17).

After half a century of informal operation by alfalfa farmers and local citizens, the Ministry of Agriculture and Water Resources established the Mezquital Irrigation District in 1945 to manage the growing irrigation system. Currently, four irrigation districts receive sewage and stormwater from Mexico City for irrigating 85,000 hectares and providing year-round livelihoods for about 400,000 people. These farmers sell all their produce back to the Mexico City area, including alfalfa, maize or corn, wheat, oats, and ornamental flowers.

No treatment of the sewage is provided, other than the natural decomposition and settling that occurs in the 60 kilometers of canals and equalizing reservoirs that transport the flow to the fields. Reliance is placed on crop restrictions to protect the public health, rather than sewage treatment. This includes bans on such crops as lettuce, cabbage, beets, carrots, radishes, spinach, and parsley.

During the 1980s, 2.2 million tons of food crops were produced annually, with a value of more than U.S. $33 million from this gravity system. Yields have been gradually increasing in the valley, probably because the shallow soils have been improved by the sewage application. The high rate of water application, due to the abundance of sewage flow, has leached soluble salts from the soils, preventing salinity problems.

Alfalfa, rotated with corn, produced several harvests during 1990, with a value of 2.3 million Mexican pesos per hectare, about U.S. $353 per hectare per year (at 6500 pesos to the dollar). The annual yields were 95 tons per hectare for several harvests of alfalfa, and 4.5 tons per hectare for the single crop of corn.

The gradual development and improvement of crop controls have prevented major health problems from occurring in the Mezquital Valley Irrigation Districts, although some problems persist with amebiasis, giardiasis, and roundworm infections, especially in children. These are also common in nearby areas without sewage irrigation, but more work is needed to prevent unnecessary increases of these diseases in the farming populations.

The sewage reuse system in Mexico City has grown and successfully expanded during the past century. In contrast, the systems in North American cities such as Boston and San Francisco have collapsed roughly every 40 years due to unsustainable designs, and have to be repeatedly rebuilt. Mexico City serves as an example for other inland cities that should irrigate farmland at lower elevations. Coastal cities, by inference, should use their domestic sewage to stimulate coastal fish and marine mammal populations.

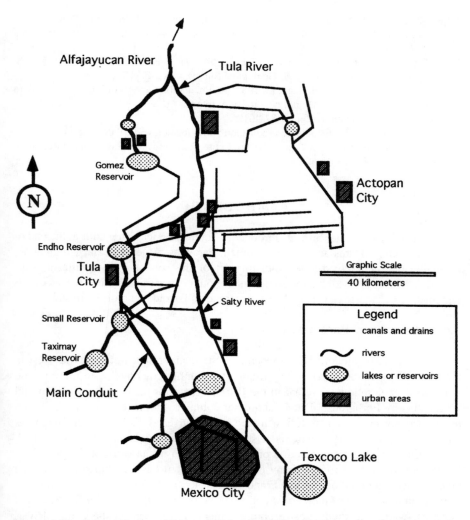

Figure 4.17 Schematic map of Tula and Alfajayucan irrigation districts. About 85,000 hectares of land are irrigated here, supplied by sewage and stormwater flows from Mexico City.

REFERENCES

Alverez, Ing H., 1993. Taller Regional Para America Sobre Aspectos De Salud, Agricultura y Ambiente, Vinculados Al Uso De Aguas Residuales. Mexico. Instituto Mexicano De Tecnologia Del Agua.

Jobin, W., Ferguson, F., and Brown, R., 1976. Ecological Review of Hydroelectric Reservoirs in Puerto Rico. Report One of the Center for Energy and Environment Research, University of Puerto Rico.

Jobin, W., Brown, R., Velez, S., and Ferguson, F., 1977. Biological Control of *Biomphalaria glabrata* in Major Reservoirs of Puerto Rico. *American Journal of Tropical Medicine and Hygiene*, 26, 1018–1024.

Peck, D., 1991. Sedimentation Survey of Lago Toa Vaca, Puerto Rico, July 1985. U.S. Department of the Interior, U.S. Geological Survey. Open File Report 90-199.

Quinones, F., Green, B., and Santiago, L., 1989. Sedimentation Surveys of Lago Loiza, Puerto Rico, July 1985. U.S. Department of the Interior, U.S. Geological Survey. Water Resources Investigations Report 87-4019.

Rivera-Martinez, J., 1979. Estudio de produccion y distribucion vertical de oxigeno en cuatro lagos de Puerto Rico, 1978. Report 9 of the Center for Energy and Environment Research, University of Puerto Rico.

4.4 THE SOUTH

There are major water projects throughout South America being built at the end of the 20th Century. Dams and canals in Brazil, Paraguay, Argentina, and Uruguay are changing the geography of the southern part of the continent in significant ways (Figure 4.18). It is hoped that the engineers of South America will watch the successes and mistakes of the engineers of North America, and avoid some of the mistakes.

4.4.1   Brazil and Paraguay

In many ways, modern Brazil resembles the frontier states of the Wild West in the U.S. Many of the romantic traditions of ranching, cowboys, gold mining, and wilderness exploration are found in the Brazilian interior far from the early coastal settlements by Portuguese, Dutch, and other colonists. In the area of dam building, federal agencies in Brazil are following the same courses as the U.S. federal agencies pursued in mid-century. There is a strong sense of "concrete fever" in Brazil, resulting in the recent construction of the largest dam in the world at Itaipu.

Before Itaipu, there were many other dams. Brazil has developed most of the hydroelectric potential on the Sao Francisco River, which flows northward; they are starting on a program of dam construction in the basin of the Amazon River, which flows eastward; and their latest accomplishments have been on the Parana River, which flows southward to Paraguay and Argentina.

4.4.1.1   Itaipu on Paraguay and Brazil Borders

The Parana River has an average flow of 293,000 cubic feet per second and divides Paraguay from Brazil. Itaipu Binacional, a joint Brazilian–Paraguayan organization poured 8 million cubic yards of concrete to dam the Parana River at Foz da Iguacu. This $9 billion project has a generating capacity of 12.6 gigawatts, the largest in the world (Figure 4.19).

It should be noted that three of the eight largest hydroelectric dams in the world are in Brazil, with a combined generating capacity over 20 gigawatts. These enormous dams — Itaipu, Paulo Afonso 1-4, and Ilha Solteira — were built by the Brazilians and some of their neighbors during the last 20 years.

Figure 4.18   The confluence of the Parana and Uruguay Rivers forms Mar de la Plata, the port of Buenos Aires and Montevideo. Itaipu Dam is the largest hydroelectric dam on earth, located on the Parana River as it leaves Brazil.

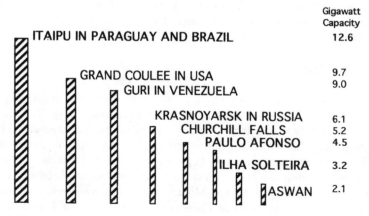

Figure 4.19   Some of the largest hydroelectric dams in the world are now in Brazil, including Itaipu, Paulo Afonso, and Ilha Solteira.

4.4.1.2 Dams in Amazonas

Tucuri Dam was constructed near Belem in 1985 on the Araguaia-Tocantins River basin at the mouth of the Amazon River, and was the largest dam to be constructed in a tropical rainforest. Shortly thereafter, Balbinas Dam was constructed on the Utuama River to supply electricity to Manaus in the Amazon interior. Balbinas has only a 111-megawatt capacity, but flooded over 2300 square kilometers of primary tropical rainforest. Recently, ELETRONORTE, the Brazilian electric authority in the north, has proposed that another massive dam be built on the Zingu River.

Environmental and Social Concerns — Despite the clear case of "concrete fever" in Brazil, and the enthusiasm for building some of the largest dams in the world, there are some negative aspects of the current rash of dam building.

It was not until 1996, after the largest dam in the world had already been built in Brazil, that the federal government established an official system for assessing the environmental, social, and economic impacts of large dams and other development projects. In retrospect, this explains the environmental and social complaints raised about many of the dams. For example, the native people along the Zingu River in Amazonas have threatened war if Brazil goes ahead with their dam plans for the year 2010.

Usually, environmental assessments have been conducted after the dams are under construction, making it unlikely that viable alternatives can be considered. Riverine people have been undercompensated for their land, and like the Cree in Canada and the three tribes in North Dakota, the government might not pay attention to their rights.

The social impact of the Tucuri Dam near Belem indicated some of the callous patterns of treating indigenous people that have prevailed in the recent past in Brazil. The dam forced the resettlement of about 15,000 people. The project encroached on three indigenous reserves, including the homelands of the Surui and Gaviao peoples. These people survive on fishing and gathering of food from the forests, as well as timber extraction and Brazil-nut production. All of these natural resources were damaged by the project, yet the impact was not evaluated until 1 year after construction on the dam began.

4.4.2 Argentina

Argentina is involved in several large water projects at present. Argentina's experience with the Yacreta Dam has been less than satisfactory. This enormous project will cost four to five times the original estimates, has large environmental and social impacts, and was still unfinished in 1996. Despite this experience, the country is eagerly embracing another large water project, but this one is a shipping canal system. The proposed Hydrovia canal system will allow ocean ships to enter the river system at Buenos Aires and Montevideo and then traverse the Parana and Paraguay Rivers upstream as far as Caceres in the center of the Brazilian state of Mato Grosso. Argentina estimates the cost for building the project to be $28 million, with $15 million required annually for dredging. Argentina opened bidding in March

1996 for dredging and placement of bouys along the Parana and lower Paraguay Rivers.

Many environmental concerns have been raised about the effect of such a long canal dredging project on wetlands in Brazil and on aquatic life. The experience of the U.S. and Canada with the St. Lawrence Seaway indicated that ocean-going ships can be important transporters of exotic species of aquatic organisms. This is one concern that should be investigated before plans for the Hydrovia canal system are finalized.

4.5 SUMMARY

The preceding chapters have summarized the situations in the Americas and illustrated the principal faults in planning and engineering modifications of rivers, lakes, and oceans. The primary impacts of human development on river basins and oceans in the last three centuries were due to:

- Dams
- Fishing, hunting, and trapping
- Urbanization and traffic
- Agriculture and deforestation
- Manufacturing
- Mining
- Navigation

Dams have had multiple negative impacts on primeval fisheries. In addition to blocking the normal passage of migratory fish, the changes they have caused in flow regimes has disrupted the seasonal patterns of stimuli that normally guide the fish as they swim up the river to spawn, and then swim down again to the ocean after maturing. The physical obstructions of the dams has also blocked the normal transport of fallen trees and branches downstream to the estuaries and coastal waters, thus eliminating an important source of organic material from the food chain of coastal organisms.

Fishing, hunting, and trapping have, until recently, operated at uncontrolled and unsustainable rates that depleted the primeval stocks of most major food fishes and animals, including whales.

The enormous growth of metropolitan areas, especially along the coastlines and major waterways, has added a new class of contaminants to our water bodies — the traffic and combustion byproducts that are washed off the streets, roofs, and paved areas of the cities by falling rain. These new contaminants include heavy metals and polycyclic aromatic hydrocarbons, as well as grease, oil, and many other petroleum products. Sand and salt used on roadways have also been found in rivers and lakes in increasing amounts. Growing human populations have also depleted groundwater reserves by increased pumping of water for domestic use. Lower groundwater storage results in lower stream flows during dry seasons.

Deforestation has proceeded rapidly in many areas, driven by the three forces of lumber production, clearing for agricultural uses, and clearing for residential developments. Although clear-cutting for lumber continues to be a major cause of deforestation, in New England and some other areas, the collapse of farming has allowed forests to recover. Major effects of deforestation are increased erosion of topsoil that ends up in rivers, lakes, reservoirs, and coastal waters, as well as loss of underground water due to the unimpeded and therefore more rapid passage of rainwater to surface streams. Irrigated agriculture has also depleted groundwater reserves by increased pumping from underlying aquifers.

Manufacturing and industrial operations have increased in size and complexity, discharging increasingly complex and sometimes toxic materials to the environment, including streams, lakes, and oceans. The list of new contaminants is as long as the list of their deleterious effects on aquatic, marine, and human life.

Mining in the Rocky Mountains and other areas of the West has in many cases left a residual of toxic wastes such as cyanide and heavy metals (e.g., lead and cadmium). These highly toxic materials find their way to streams and rivers, either through untreated mine drainage systems, through failure of waste storage dams, or after closure of mines when seepage overflows protective berms or dams.

Navigation channels such as the St. Lawrence Seaway have allowed foreign species of migratory eels, fish, and mollusks to penetrate large freshwater systems, using the same route as the ships that ply the canals, and often attaching themselves to the ships to obtain free rides. Biological analyses of the potential for introduction of foreign or dangerous species have not been part of shipping canal design in the past.

It is clear that Brazilian and other South American engineers and water planners are having a long episode of "concrete fever," and it is difficult to say what influence environmental or human rights pleas will have. But one can hope that these highly competent engineers will take advantage of the experience of dam builders in North America, and try to make their projects sustainable, environmentally beneficial, and socially responsible.

REFERENCES

Barrow, C., 1987. Environmental Impacts of the Tucuri Dam on the Middle and Lower Tocantins River Basin, Brazil. *Regulated Rivers: Research & Management,* 1(1), 49–60.

Bermann, C., 1995. Self-managed Resettlement — A Case Study: The Ita Dam in Southern Brazil, *International Journal of Hydropower and Dams.* June 1995, p. 149.

Goodland, R., Juras, A., and Pachauri, R., 1993. Can Hydro-reservoirs in Tropical Moist Forests Be Environmentally Sustainable? *Environmental Conservation,* Summer 1993, 20(2), 122.

Gribel, R., 1990. The Balbina Disaster: The Need to Ask Why? *Ecologist,* July–Aug. 1990, 20(4), 133.

Hasson, G.D., Suarez, C.E., and Pistonesi, H., 1995. Energy and Environment in Argentina. Past and Prospective Evolution. United Nations Environment Programme, Roskilde (Denmark). Collaborating Centre on Energy and Environment. April, 179 p.

Heath, R., 1995. Hell's Highway. *New Scientist,* 146(1980), 22–25.

Itaipu, 1990. The Rock That Sings. *Eau, Industrie, Nuisances,* 1990(140), 121–123.

Martin, E.J. 1989. Definitional Mission for Feasibility Study for the Canals of Martin Garcia, Uruguay, S.A.: Feasibility Study for Environmental Projects. December 1989, 6 p.

Monosowski, E., 1991. Dams and Sustainable Development in Brazilian Amazonia. *Water Power and Dam Construction,* May 1991, 43(5), 53.

Pearce, F., 1995. The Biggest Dam in the World. *New Scientist,* 145(1962), 25–29.

Power, G., 1985. Will Tucurui Dam Damage Amazon Ecology? *World Water,* 8(1), 37–38.

World Water, 1983. Power on the Parana Boosts Brazil's Potential, 6(10), 17.

Our Hope for the Future

Previous sections contained descriptions of recent ecologic conditions in major hydrologic systems of North and South America. Unfortunately, the descriptions included declining water quality, disappearing stocks of whales and fish, and large engineering projects that were poorly tailored to long-term water needs of growing human populations. The sustainable societies and ecology of the *original* Americans have been displaced by extractive and unsustainable activities of the highly productive *immigrant* Americans who have dominated the Western Hemisphere for the past few centuries. This recent history of the Americas uncovered a clear need for a synthesis of these two cultures — a productive and sustainable society to be formed by *new* Americans.

This section proposes an approach for water management by the *new* Americans, to be led by citizen groups organized along watershed geography. These groups should take over the direction of the ecological restoration of the Americas in the face of the recent collapse of governmental programs. A new approach to the education of water planners and design engineers is also put forward, to provide sustainable infrastructure for the *new* American society. Planning of the *new* America should be based on ecological principles rather than on short-term economic analyses.

CHAPTER 5

Citizen Watershed Associations

The Green Movement from 1970 to 1995, though ephemeral, was perhaps the most successful of the periodic sanitary awakenings that have swept over countries in their struggles with the exponential increase of waste products accompanying their population growth. It was also worldwide. In the first decade while the movement was confined to cleaning up water in the Americas, it was quite successful, easily overcoming corporate and municipal resistance to remediation because of well-established traditions and laws related to clean water. But when the Green Movement expanded to air and land pollution, it trod on the toes of big industry — a fatal offense.

The reaction against the Green Movement in the U.S. began in earnest with the federal elections of 1992 when industrial funding for conservative politicians and platforms exploded. Two years later, the conservative revolution in the U.S. Congress began to dismantle the federal environmental program, and by 1995 the Green Movement was dead. The global demise of the Green Movement was arranged by the multinational corporations that wanted no more interference in their industrial production of chemicals, extraction of coal and oil deposits, or industrial harvesting of fish from the oceans.

Usually, the great sanitary awakenings tapered off because of their own success in eliminating the environmental damage that spawned them. As the contamination receded from public consciousness, the fervor of the movements abated. But in the Green Movement, this normal ending of a social phenomenon was accelerated by the strong and well-financed political resistance from all manners of corporations. The Movement had overstepped its power base and was not well-enough organized politically to handle the reaction.

If our environment is to be protected in a sustainable manner, we have to be aware of this political reality. We need to turn to another social organization, one that cannot be purchased by wealthy corporations. Some of the more successful aspects of the Green Movement may shine some light in the direction of a more sustainable course that we can follow. The experience of the Neponset River Initiative and the Nashua River Watershed Association in Massachusetts can take us in one of these potentially valuable directions.

The plaintive and unsuccessful attempts at protecting our environment through government programs began with creation of the Metropolitan Sewerage Commission in 1889 for protection of Boston Harbor, followed by the Metropolitan District Commission after World War II, and most recently culminating in the formation of the Massachusetts Water Resources Authority. Because of their narrow engineering philosophy and their organization along political boundaries instead of watersheds, these agencies were never able to overcome the inertia of their concrete and steel solutions, and could not come up with innovative or ecologically sound approaches. Conservation and reuse of water, wastes, and resources were never given serious attention by these carefully trained engineers.

To break out of this repeating trap, shoreline dwellers must form themselves into communities or watershed associations to plan and implement the rehabilitation of their rivers and restore the primeval fish populations. The process has already begun in Massachusetts, where over 100 river and lake associations have organized. Their initial focus has been to monitor the conditions in the rivers, then to assist government in improving the rivers. Some of the associations have forged ahead and obtained funds for purchasing riparian lands and for public education. Many of them have been active politically and have assisted in the passage of bond issues to be used for further restoration.

5.1 MASSACHUSETTS WATERSHED ASSOCIATIONS

Fortunately, these community-based organizations in Massachusetts, organized by the basic resource unit of the watershed, and including the users of the resource as well as its polluters, have taken very creative approaches. These groups include fish and wildlife enthusiasts, owners of riverside and lakeside homes, and a large number of women with strong stewardship instincts. Most of the strong watershed associations in Massachusetts are headed by women, the most famous being the Nashua River Watershed Association. This group is responsible for the complete rehabilitation of a river that once ran the color of whatever type of paper was being produced in the Nashua and Fitchburg mills that day. The Charles River Watershed Association and the Neponset River Watershed Association are also headed by women, with similarly successful histories.

Much of the land along the Neponset River has been given extra protection against encroachment, and the Neponset River Watershed Association was instrumental in assisting passage of an Open Spaces Bond Issue, which is being used to permanently set aside the estuary and other sensitive parts of the floodplain for wildlife protection. The beauty of this approach is that the land values had been depressed by past industrial contamination, so prices were extremely low in the late 1990s. It is hoped that these watershed associations will grow in scope and effectiveness as the government programs of the Green Revolution sink into oblivion.

The existing watershed and lake associations in Massachusetts should continue to expand their base, and also to group into larger units. For example, the Charles River and Neponset River Associations could join with Save the Harbor, Save the Bay to form a rural, suburban, and urban alliance of all residents of the Boston

Harbor drainage area, and perhaps eventually the Massachusetts Bay drainage area. As long as the groups are solidly founded and related in a democratic and just alliance, they should be successful. They might pattern their association on those of the Six Nation Confederation of the Iroquois tribes.

5.1.1 Focus on Massachusetts Bay

There are many possible courses of action for these citizen groups, but it is hoped that those in Massachusetts will focus on restoration of the waters of Boston Harbor and Massachusetts Bay, with a program for rapid restoration of fish and whale populations in rivers, estuaries, and coastal waters. The associations should then broaden their reach to all of New England and to the Pacific Northwest, to concentrate on the restoration of salmon fisheries.

It is important that such associations also establish legal and governmental recognition, and develop plans to derive revenues from the resources they improve. For example, they might develop ecologically sound hydropower installations for low head dams. As the efforts of the citizen groups improve the ecology and biodiversity in the watersheds, the groups have a right to claim a portion of the increased state revenues from fishing, clamming, and hunting activities. Perhaps they could also share in the increased property values of riverside land by investing in land now when its value is low, and selling it after they have raised its value.

The watershed associations should be generally inclusive, but exclude those commercial and industrial concerns that have been abusing the rivers for the past century. Sportsmen, naturalists, citizens groups, towns and cities, shoreline landowners, and public leaders should all be enlisted to develop these watershed associations into self-sufficient organizations. However, if they include the water-abusing industries that have caused the problems in these rivers from the beginning, the more powerful corporations will use the watershed associations for their own purposes, which are commercial advantage and profit. In this case, the rivers will simply revert to the foul, stinking, and dead state in which we found them before the first Earth Day.

5.1.2 Water Abusers

A distinction should be carefully drawn between water-abusing industries and other shoreline landowners. The industries have destroyed the real estate values of land along the rivers and ponds by damming the rivers, polluting them, and then dumping trash and industrial wastes along the banks. However, the other landowners are the ones who have suffered from this desecration.

While the industries were making a profit at the expense of the rivers, the landowners were losing money on their investments in property and homes. What may have originally been a rustic summer cabin on a beautiful lake often became a financial liability as the foul summer odors made it so unattractive that people preferred to drive 50 miles to the ocean instead of swimming in what used to be the cool refreshment of the primeval waters.

The legacy of industrial rape will persist for another generation, even if it has been stopped. The industries will use every means possible to avoid paying for remedial activities that would reduce their profits. As a result, enforcement activities by state and federal agencies against these industries have been delayed and diverted for years, and little progress should be expected. But perhaps there is still a solution. It must come from the watershed associations as they grow in strength.

The watershed associations should become stockholders in those industries that are responsible for the remaining problems on the rivers, and should then redirect the industries to take responsibility for the effect of their previous profit-taking. In this way, the profit will be returned to the rivers, from where it was originally extracted.

5.2 A NEW PROGRAM FOR THE NEW AMERICANS

Many industrially developed rivers and important harbors in the Americas are at an important nadir in their value at the end of the 20th Century; thus, land and water rights should be purchased now by the public for future restoration in the 21st Century.

It is also time for the environmental movement to shift emphasis from negative restrictions on economic development in the U.S., to a positive program to stimulate the economy while restoring the environment at the same time.

Major economic damage was caused by mindless industrial development and commercial fishing in the 19th and 20th Centuries, which destroyed freshwater, coastal, and ocean fishing in North America and elsewhere. This destruction came about through rapacious over-fishing, development of numerous small dams on the mainstems of coastal rivers, and the discharge of toxic industrial wastes to these same rivers, harbors, and bays. The damage extended to the ocean, where over-fishing and whale hunting reduced marine populations to disastrous levels. In the 19th Century, the freshwater fishing industries of New England collapsed. During the course of the 20th Century, the whaling industry collapsed; and in the course of the last decade, the marine fishing industry of New England collapsed. The negative impact of this ecologic collapse on New England and the Pacific Northwest has been a major economic disaster.

Most of the industrial dams are now abandoned, and there are no longer any excuses for toxic discharges from industries that should be recycling their prime materials. Thus, the rivers and coastal waters destroyed in the last two centuries can now be reclaimed. The process must begin in the coming century by renovating these dams and cleaning up the discharges, including toxic sediments.

In the 21st Century, the reclamation of our waters must begin by specifically restoring fishing as a major economic activity in New England, the Pacific Northwest, and other coastal areas of North America. This could also be undertaken as a major national program in Canada, the U.S., Mexico, and the Caribbean Islands.

There is now adequate understanding of fisheries and aquatic and marine ecology to develop this as a major economic effort, with large benefits. Sea Grant colleges in the U.S. should become major foci for developing these programs in the coming century, just as Land Grant colleges helped to develop our enormous agricultural production in the past century.

CHAPTER 6

Dam Designers and Water Planners

This chapter summarizes major mistakes included in many water projects in the past, and then covers elements needed by the *new* Americans for sound water planning and sound engineering design.

6.1 MISTAKES

In addition to all of their intended positive effects, human development activities have had the following detrimental impacts on American water resources during the last three centuries. These negative impacts and mistakes must be addressed in the broad planning as well as in the detailed design of dams, canals, and sewers in the new America.

6.1.1 Dams

The principal negative ecological impacts of dams included decreases in the amount of suspended solids, including terrestrial organic material reaching estuaries and the ocean, and thus food for zooplankton and carnivore fish. They also caused temperature changes that drastically affected fish habitats. The dams and reservoirs slowed flood waves when empty, but caused local flooding when full. Dams blocked the normal passage of migratory fish, and in addition to eliminating the flowing habitats flooded by the reservoir, the ecology of the flow downstream was also changed drastically.

6.1.2 Agriculture

A basic requirement for expanding agriculture was to cut forests and under-growth, which increased the amount of runoff. Application of animal wastes and industrial fertilizers to land increased nutrients in water.

6.1.3 Deforestation

The widespread cutting of forests decreased the rate of delivery of driftwood to estuaries and oceans because the amount of fallen wood in streams was reduced. This also reduced the habitat needed in tributaries for spawning. Programs for clearing of streams to avoid flooding further reduced habitats and transport of decomposing wood to coastlines. Elimination of the wood, as well as excessive trapping for pelts, nearly exterminated the beaver populations. Their return would do more to restore wetlands to North America than any conceivable government program.

6.1.4 Urbanization

Increased flows of domestic sewage caused an increase of dissolved nutrients in waters and thus algae and herbivorous fish. The increased area of pavement raised the amount of runoff, and storm sewers sped runoff and increased contaminant load to natural drainage systems. The engineers channelized normally meandering and flooding streams, decreasing the extent of slowly flowing and protected habitats for aquatic organisms.

6.1.5 Transportation Systems

Increases in the area of pavement and amount of urban and long-distance traffic changed the nature of the material in runoff from normal organic material that could be utilized by aquatic and marine organisms, into particulate and toxic materials. The amount of sand and salt increased, while new classes of hydrocarbons were added (e.g., gasoline, diesel, grease, and oil). Heavy metals from engines and bearings were washed off the roads and accumulated in predators, long-lived aquatic animals. The principal damaging metals were lead, zinc, and copper.

6.1.6 Manufacturing

Since the colonial era, power for manufacturing and general industry was provided by construction of dams and water mills. In time, the reservoirs behind these dams were used for process water and for dilution of industrial wastes returned to the river downstream of the dam. A principal damaging impact of these systems was the discharge of the wastewaters to natural stream systems. The industrial wastes included heavy metals, industrial solvents, as well as organic material that consumed oxygen. In addition, all kinds of industrial rubbish were discarded along river shores. Because of air pollution by industries, there was also a significant increase in fallout from air contamination.

6.1.7 Mining

The short-term and often poorly organized nature of many mining operations caused large amounts of waste runoff from mined materials and process chemicals. In addition, the primitive systems developed for mine drainage often used the natural drainage system to absorb toxic materials. Mine operations often ended in bank-ruptcy when the resource was depleted, leaving rusting and decaying hulks of structures harboring hazardous and toxic wastes.

6.1.8 Navigation

The channelization required for passage of large watercraft often eliminated normal floodplain dynamics, and allowed passage of exotic organisms that had previously been excluded by natural hydraulic obstacles such as high waterfalls. Many exotic organisms have also been introduced into the freshwater systems of the Americas from other continents, usually in the bilge waters of ocean-going ships.

6.2 FAULTY ENVIRONMENTAL DESIGN CONCEPTS

The detrimental impacts of water projects that severely damaged our rivers, lakes, and oceans were due to faulty design concepts used by American dam and water engineers. These design concepts should be corrected. However, the engineering mistakes were in the context of a society focused on extractive enthusiasm, engineering simplification, and ecological disaster.

The extractive enthusiasm of the immigrant American society ranged from trapping of beavers in the West to drilling for oil in the Southwest, shooting buffalo in the West, mining coal in Appalachia, clear-cutting in the northern Atlantic and Pacific forests, whaling in the global oceans, and otter-trawl fishing on Georges Bank in New England. No one seemed to learn from the disastrous effects of these extractive activities, and as soon as one resource was depleted, the extractors were off to another. Sustainable harvesting was never developed and the "tragedy of the American commons" continued unabated.

The simplifying tendency of engineers made it possible for them to attempt enormously complex tasks with simple theories and calculations. They built arch bridges, railroads, dams, and canals on every kind of terrain, using the same basic equations. In a short time, they covered the continent with their structures. The enthusiastic extractors who needed railroads to haul their oil, bridges to get to their mines, and ships and trawls to bring in their fish and whales were pleased with the prodigious production of these engineers. However, this same trait also meant that many of the engineers' creations failed when the simplifying assumptions proved false.

With dams, the failures were usually not obvious for many years, especially when the simplification concerned the character of the river that had been dammed. Bonneville Dam on the Columbia River was built without regard for the salmon in the river because it was assumed that they were unimportant. In the economic cost-benefit for the dam, no value was assigned to the loss of the salmon passage, despite the fact that 100 million pounds per year of salmon may have been harvested at one time from the river. At a current market value of $5 per pound, that would be an annual income worth half a billion dollars. Perhaps simplifying assumptions led those engineers to err.

6.2.1   Dams and Canals

Dams were designed with single, usually industrially oriented purposes, thereby neglecting and usually impeding development of several other important functions of the river in the ecology of the basin. Dams and canals were designed without regard to primeval aquatic and marine life, using concepts derived from oversimplified analyses based on experiments in laboratory canals and tanks.

The use of short project lives was a common mistake in the design of large projects, without providing sustainable strategies for operation or for decommissioning. Dams were constructed that eliminated food supplies in estuaries by blocking downstream transport of woody debris and other organic material. River hydrology was drastically altered by dams that dampened necessary flow stimuli for migratory fish.

6.2.2   Programs to Control Environmental Contamination

Enforcement activities in the early days of the Clean Water Act were based on simple chemical indications of contamination, expressed as water quality standards, instead of using broad ecological assessment of fisheries and other populations at the top of the ecological food web. A major omission in most water project analyses was neglect of the economic impact of water pollution on land values. Most economic analyses disregarded historical perspective and the need for a sustainable future of water projects. Ecological analysis of river contamination was conducted without regard for the important role of seasonal changes, and of large storms and hurricanes. Naive initial attempts in environmental enforcement were made to control industrial pollution without regard for the social and economic influence of wealthy industrial corporations. Natural resource and environmental agencies allowed uncontrolled harvesting of naturally limited populations of fish and mammals, leading to their destruction.

6.2.3   Sewer Design

Sewers were designed for the water carriage system of waste disposal, using expensive pure water supplies to transport wastes to rivers, lakes, and oceans. The sewerage systems allowed mixing of storm runoff, as well as industrial and human waste streams, instead of maintaining separate resource and waste streams to facilitate their reuse. A major design flaw in sewerage systems was the use of water bodies and oceans for repositories of nondegradable wastes, especially heavy metals and toxic industrial chemicals. Urban water and sewers were designed with disregard for the patterns of growth limitation observed in smaller organisms that contaminate their water and environment, as well as depleting their food supplies. Urban sewerage systems were designed that required unsustainable operation and maintenance efforts. Storm sewer systems were aimed at speeding transport of runoff to natural streams, instead of retarding the runoff to preserve the hydrology of natural drainage systems.

6.3 SOLUTIONS

To correct the mistakes of the past three centuries, American water planners and engineers should revise their design methodology. There is an urgent need for this because the growing human population will soon force the construction of new dams in order to increase food production and domestic water supplies.

6.3.1 Yes, More Dams

To meet the future needs for increased grain production, there will soon be a need to build even more dams in order to stabilize the food supply and to link it to population growth.

In drought-prone areas, dams will need to be large enough to store water from a series of wet years through the end of a series of dry years, based on regional hydrology and adjusted for apparent changes in future weather patterns expected from global warming. Thus, large dams should be more common in the future, not less. The need for over-year storage is due to the increasing severity of weather cycles, and the increasing dependence on local grain production for food, as human populations exceed the food supply, now reduced because of the loss of fish and other foods from the sea.

In areas of erratic rainfall such as New England, the local food supply from most crops is limited by minimum rainfall. With over-year storage, this can be raised to the level supported by the mean rainfall. As long as the surrounding industrial communities use water carriage for excreta disposal, the addition of recycled domestic sewage for irrigated agriculture can also be used for irrigation, and thus the food supply can rise as the population rises.

In regions continuously threatened by drought and close to famine, such as the Sahel region of Africa and southern Africa, inter-year storage in dams is a necessity for survival through the biblically long 7-year droughts. If weather changes create similar problems in the Americas, more dams and larger dams will be needed in these areas.

About 1000 dams were built along the New England coastal hills since the middle of the 19th Century. Most of them are now over 100 years old, and most of them have also lost their original function. The corporations and even the industries for which many of them were built have disappeared, leaving an unwanted legacy that is wreaking havoc with the primeval river ecology, without compensating benefits.

Abandoned dams and their sediment-trapping reservoirs are local sources of ecological damage. The concentrations of heavy metals, PCBs, and other toxic industrial wastes are so high in the sediments of many of these dams that disposal or detoxification has become a monumental task. Most of the dams are now slowly bleeding toxic materials into the rivers and oceans, in an uncontrolled process that most regulatory agencies would prefer to ignore.

A rational policy of operation and modification must be developed for these old dams, and rational designs must be implemented for new dams to avoid repetition of the problems. They will not go away by themselves.

6.3.2   Conversion of Single-Purpose Dams

Because of the inexorable rise of food as a primary human need, multipurpose dams that cannot be removed or modified must eventually be converted to an almost exclusive emphasis on food production, whether it be for irrigated agriculture or for fish, mollusks, crustaceans, and marine mammals. Hydroelectric power could be generated, as long as the water would not be lost from food production.

6.3.3   Irrigation Dams and Canals

A good example of the way to improve irrigation systems in North America is the potential use of domestic sewage from Denver, Colorado, for irrigation of the flat land along the South Platte River.

Denver's domestic sewage should first be separated from industrial and other toxic wastes, and then used for irrigation after simple treatment, rather than giving it extensive and expensive treatment and thereafter discharging to surface or ground-waters directly. This would slow the demand for more irrigation dams, especially some of the more ridiculous dams recently proposed for the South Platte River near the Colorado–Nebraska border.

Use of Denver's sewage for irrigation would also make it possible to free some of the rainfall on the east slope for domestic water supply, instead of applying it as irrigation water.

6.4 ECOLOGICAL PRINCIPLES FOR WATER PLANNERS

Planning of water projects in the new Americas should include the following five major elements, which also become the new basis for review of curricula in water planning and water engineering courses in colleges and universities. It is time for a broader approach to engineering education regarding dams and water resources.

6.4.1   Designs Based on Sustainable Methodology

Realistic projections of human population growth should be made in order to determine the sustainable populations for the available water. If the proposed dam cannot be designed to meet long-term needs, it should be designed with a decommissioning plan. Dams should be built for multiple uses that will allow increasing emphasis on irrigation.

6.4.2   Basic Ecological Principles of Sustainable Systems

Five of the most easily identified ecological principles include:

1. All water resources and water projects should be utilized for multiple purposes, and none should be designed for narrowly conceived, single purposes.
2. Industrial materials and water should be recycled.

3. There should be zero industrial discharges to the environment, even to sewers.
4. Design of domestic water supply systems should be highly conserving.
5. Sewage systems should be designed to recycle domestic water and solids.

6.4.3   Historical Perspective

Dam, canal, and sewer designers need historical perspectives of both the past and the future. An understanding of past efforts of the community will help avoid repeating the same mistakes. A long look into the future is needed to determine truly sustainable solutions.

6.4.4   Natural History of Aquatic and Marine Life

Water engineers should understand the natural history of the fish, mollusks, crustacea, and other organisms that presently inhabit the bodies of water under consideration. The engineers should also know the life stories of riparian and marine mammals, as well as the basic elements and ecology of the forests, aquatic vegetation, and algae that support these organisms.

6.4.5   Restoration of Waters by Focusing on Fisheries

Water planners should begin to focus on restoration of the aquatic and marine resources of water bodies, with the intent of restoring populations that can be harvested on a sustainable basis for food supply. These populations of animals could range from snails to whales.

The potential for riverside farming should also be included in these plans, especially when sewage discharges occur near flatlands with a favorable climate for agriculture. The sewage can be used to irrigate animal feed crops, grazing land for deer and other large mammals, and perhaps also for aquaculture.

Some of the specific ecological solutions proposed here are to:

- Remove the most destructive of the dams, while retaining those that can be used to restore primeval hydrology.
- Reforest river banks and watersheds.
- Reconfigure storm drainage to hold water in upper watersheds.
- Sustainably operate sewage treatment systems, such as the one in Boston Harbor. The Boston Harbor example shows that separation of resource and waste streams, and basic plumbing, are more important than expensive, advanced treatment facilities that have been shown historically to be financially unsustainable.

6.5 ECONOMIC VALUES IN RESTORATION OF FISHERIES

Based on simple economic considerations, there should be significant economic advantages in restoring the ecology of most American rivers and oceans. The consequent renaissance of fishing, hunting, and trapping would quickly restore one of the fundamental and basic sectors of the New England economy, and should be effective in other parts of the country as well.

The focus for restoration of our aquatic and marine resources for the coming century should be the reestablishment of the primeval fisheries that sustained the original Americans. In New England and the Pacific Northwest, this means restoration of salmon and sturgeon, as well as shad, menhaden, herring, and smelt, and the reestablishment and protection of the beds for mollusks and crustacea.

This will cost money, but should be seen as a long-term investment in the economy and in the diversity of our food sources. The diversity is not only nutritionally and economically important and would add flavor to our lives, but it would give a robustness to our food supply in the face of the unpredictable vagaries of climate and industrial disasters.

Restoration plans for rivers and bays should focus on the ecologic needs of these fish, mollusks, and crustacea. Renovation, modification, or elimination of dams, aqueducts, and drains should be based on the ecological needs of these aquatic and marine resources.

Within that framework, programs for rehabilitation of the old industrial dams of the Atlantic coast should evaluate the potential of each of these myriad low dams for:

- Installation of effective bypass systems for fish
- Elimination of ineffective fish ladders
- Operation for fish and recreation
- Modification to recreate primeval hydrology
- Hydroelectric generation using low-head technology favorable to fish
- Removal of the dam

The economic returns that should be included in cost-benefit assessments of these programs are:

- Low-head hydroelectric power
- Restored fisheries of major rivers, estuaries, and bays to original state
- Restored and sustainable harvests of fish, mollusks, and crustacea
- Increased recreational fishing
- Increased riparian property values
- Increased real estate values of riverbank buildings
  (There are important examples in the abandoned mill and related dam sites in East Walpole and Norwood in Massachusetts along the Neponset River. Industrial pollution of Neponset Reservoir was shown to cause direct losses of property values approaching $15 million for even a small 300-acre reservoir. Indirect losses to the surrounding populations in recreational and esthetic values probably double or triple this figure.)
- Reduced disposal costs for contaminated sediments dredged from harbors, such as those in Boston and New York

The Clean Water Acts of the 1960s and subsequent decades defined the chemical and bacterial water quality required for fish populations and for other uses such as swimming. Although there has been significant progress in improving the basic parameters of oxygen and gross chemical quality needed by fish, many additional problems have impeded development of fisheries and other public uses. While slowly improving water quality, over-fishing, habitat destruction and watershed modifications have been destroying the basic fish populations in our lakes and rivers, and even in the oceans. A major integrated effort is thus needed to restore our original fisheries, based on our best understanding of aquatic and marine ecology.

On East Coast rivers, many dams have outlived their purpose and have now become ecological liabilities because they trap wastes and toxic sediments. The dams have also adversely modified the normal hydrology of the rivers, reducing nutrient loads to wetlands, estuaries, and coastal fisheries.

Filling of coastal and riparian wetlands has diminished the size of potential fish and shellfish habitats, but there is little that can be done to remedy this loss. Dams will also be difficult to improve, but the smaller industrial dams are the most likely candidates.

The potential value of rivers, reservoirs, and estuaries for food production and other economic benefits need to be newly evaluated to determine potential costs and benefits of ecologically based modifications to existing dams. Owners of existing dams should be responsible for their rehabilitation. New dams must be designed to optimize all uses of the water in a sustainable manner.

6.6 METHODOLOGY FOR ESTIMATING WILD SUSTAINABLE HARVEST

A concept that should be included in the economic analysis of proposed water projects or rehabilitation of existing systems is the estimation of a sustainable harvest from the water body involved. The maximum value for this sustainable harvest would be equivalent to the harvests available to original Americans, such as the salmon and sturgeon harvests of the Columbia River tribes at the time of the Lewis and Clark Expedition.

This estimation of a sustainable harvest must also include the principles of stewardship and dependency. Stewardship requires a reverence for all life, and would lead people to hunt, trap, and fish only for food and clothing. Instead of hunting for sport alone, people might become trackers and photographers to develop preservation strategies for the species at risk.

The estimation of sustainable yields from restored fisheries can begin with the estimation of values for yields per acre from currently unspoiled areas, multiplied by the acres of habitat potentially available in currently damaged areas.

- For rivers, multiply shoreline by typical habitat cross-section, either width of strip or volume
- For lakes, use shallow margin areas
- For wetlands, use surface area

The species that could be harvested include a wide range of animals and even plants (Table 6.1). The enormous potential for sustainable harvesting of these species has seldom been recognized, and should revolutionize economic analysis of proposed water projects.

Table 6.1   Potential List of Animals and Plants that Could Be Harvested for Valuable Food and Fiber in Restored River, Lake, and Coastal Habitats

Species	Information requirements for estimations of sustainable populations and harvests	
1. Marine mammals		
Whales	Feeding areas	
Dolphins		
Porpoises		
Seals		
2. Finfish	Spawning area	Feeding area
Freshwater		
Anadramous		
Catadramous		
Coastal		
Pelagic		
3. Shellfish	Beds	Fish for life cycle
Marine		
Coastal		
Freshwater		
4. Lobsters	Feeding area	
5. Crayfish		
6. Shrimp		
7. Crabs	Feeding area	Hiding places
8. Terrestrial mammals	Wetland habitat areas	
Food		
Deer		
Fur		
Muskrat		
Mink		
Beaver		
Otter		
9. Birds		
Pheasant		
Ducks		
10. Trees and vegetation		
Fruits		
Berries		

6.7 BIOLOGY OF MIGRATORY FISH

Another major requirement for water planners is to understand the lifecycle and biology of the major migratory fish, especially salmon, sturgeon, shad, and herring. This information is a necessary element in the design of large water projects that will support these fish populations. A great deal of recent information has been developed for the Columbia River and Connecticut River fisheries, which is invaluable for planners.

6.8 GUIDELINES FOR DAM ENGINEERS

Engineers who develop detailed designs for the dams, sewage systems, and other structures conceived by the water planners must use new design concepts. These new engineering design concepts include the need for dams that can be ecologically operated for multiple purposes. Such operation requires a basic appreciation of the ecological needs of the watershed and the organisms that previously inhabited it.

The simple and mechanistic designs of most industrial and hydroelectric dams in the past were based on industrial needs, often completely violating the ecological and basic human needs of riverine and downstream communities. The design and operation of dams in the future must be based on maintaining the original ecology of the riparian and downstream populations.

In simplest terms, those dams such as Glen Canyon Dam on the Colorado River must be designed and operated to release flow in a pattern that mimics the original seasonal pattern of floods in the spring and dry streambeds in the late summer.

This design requirement will shift the benefits from the new dams to a more equitable distribution between the urban and riparian human populations, and to the primeval populations of trees, fish, mollusks, and whales that originally inhabited the rivers and coastline downstream of the dam sites.

6.8.1   Retain Natural Seasonal Pattern of Flow

Dams should retain or restore principal features of natural hydrology and ecology, to pass spring or hurricane flows and flush out sediments. They should be designed, operated, and modified to optimize use of all aquatic resources. A bold experiment to modify operation of Glen Canyon Dam on the Colorado River is exploring the impact of recreating the pre-dam spring flood downstream. In March 1996, an artificial flood of 45,000 cubic feet per second was passed under the dam for 2 weeks, as a modest simulation of the primeval torrents from melting snow that shaped the Grand Canyon over past millennia (Figure 6.1). Observations on erosion, vegetation, and fisheries were made downstream during the following year, and a second artificial spring flood was released in 1997. As a minimum, these floods will help restore the primeval fisheries that lived in the Grand Canyon, They might also reestablish the vegetation and sand banks that made the canyon such a paradise for explorers and whitewater rafters.

Construction of dams along coastal streams has caused an important and largely unrecognized curtailment of the normal process of erosion and washout of soil, vegetation, and woody debris, which previously enriched downstream swamps, estuaries, and coastal fisheries. Not only do the dams provide a physical obstacle to the transport of this organic material, but the wood is frequently collected from the dams and burned, thus taking it completely out of the aquatic cycle.

In attempts to minimize flooding, many states provide funds to assist local governments in clearing fallen trees and debris from small streams. Unfortunately, this process also deprives the downstream aquatic communities of their food supply. Woody debris and fallen trees are important items in the aquatic ecology of the rivers as they provide local conditions in streams for spawning of herring and salmon, as well as being a source of nutrients for all organisms.

Figure 6.1   Glen Canyon Dam in March 1996, showing experimental discharge of 45,000 cubic meters per second. The experiment is an attempt to recreate primeval ecology of the Colorado River as it passes through the upper portion of the Grand Canyon in Arizona.

When the trees or branches rot under attack by insects and fungus, the detritus is an important source of carbon for the food chain of the aquatic and marine organisms downstream. Prior to construction of dams along the coast, the estuaries and coastal waters received continual infusions of large numbers of insects and wood-boring mollusks in this decaying wood as it was washed down by spring floods and summer hurricanes.

Studies of tuna fish in the Pacific Ocean showed that offshore drifts of this rotting wood were also an important focus for schools of tuna in the open ocean. The tuna were sheltered under the wood, perhaps for food, shade, or other reasons.

Dams on the tributaries and rivers of the coast have a double impact on anadramous or migratory fish. They deprive them of the carbon in their food supply as well as block them physically from their annual migrations, which are necessary for reproduction. Little wonder that anadramous fish have practically disappeared from both the Atlantic and Pacific coasts.

6.8.2 Seasonal Passage of Accumulated Algae and Nutrients

In New England and along the East Coast Piedmont, dams should be designed to allow the operator to dump water from the reservoir quickly in early fall before the algae die. Also, low-level outlets are needed to pass dead algae as they settle to the bottom. Operators should shorten reservoir detention times by recirculating flow through simple sand filters using gravity flow — or maybe through closed pressure filters to minimize head loss for recirculation.

6.8.3 Instream Treatment

Abandoned industrial facilities and government properties along rivers should be utilized as passive instream treatment facilities such as tube settlers, sand filters, settling basins, and nurseries for aquatic life. This property is currently of low value because of deteriorated condition of water bodies, and is at an opportune price for confiscation by public agencies as compensation for industrial damages to the watershed. Another approach would be to suggest that industries donate land and facilities to avoid fines and prosecution.

6.8.4 Protection and Passage of Fish

Dams should be designed with the recognition that fish and other organisms in their reservoirs need special protection:

- Fish ladders do not work; thus, other solutions are needed, such as conduits through dam bases.
- Dams should be supplemented with fish and mollusk nurseries.
- Oxygenation may be needed for protection of fish in low-flow conditions.
- Dams and the drainage systems around them must be able to handle the impacts from road salts, sand, and road debris, and from storm sewer flows.

6.8.5 Zero Discharge Policy

Instead of discharging organic and industrial wastes into surface waters, we should recycle them:

- Separate industrial, domestic, and storm wastes
- Domestic wastes should be used for sewage irrigation
- Industrial wastes should be recycled or compacted
- Storm waste should be settled and clean water released

6.8.6 Project Life — How Long?

The design estimation of the lifetime of proposed dams must be reconsidered in the light of over 100 years of experience with modern dams. There are two aspects to developing more rational design lives for dams. One is that the original dam must be built with provisions for its decommissioning, perhaps even its removal. The second is that the benefit and cost analyses must include a more realistic estimate of lifespan, as well as the consequences of decommissioning.

An exit strategy is needed for industrial or mining dams, with provisions for the end of useful life of the dam or mine. A good place to start would be with the old industrial dams on the East Coast Piedmont.

Most of these are a century old, are no longer used, and cause great damage to the fisheries and other aquatic life that originally existed abundantly and profusely in these rivers and estuaries. One by one, the river systems — including mainstem, tributaries, estuaries, and bays — should be analyzed to determine the necessary steps to restore the aquatic and marine food resources to their original health. The dams should be designed, modified, and operated to deal with problems of excess algae and toxic sediments.

Most of the solutions will involve changes in operation, modification, or removal of the old dams. This is the responsibility of owners and they should be required to do so, removing toxic sediments in a safe manner before the dams are removed. The economic benefits from these programs may more than justify the expenses.

CHAPTER 7

The End of it All

Three major themes were developed in this book regarding the design and construction of dams and water projects in the Americas. The first theme was the need to build new dams and to rehabilitate old dams for ecologically sound operations that meet the growing needs of our expanding populations. This is especially important in the face of the depletion of our uncontaminated reserves of water for public use.

The second theme was the need to radically change the educational framework for dam designers and water resource planners to teach them to develop sustainable projects, adaptable enough to changes in human communities and the environment so that they will be self-sustaining for generations. This will require modification and broadening of current engineering curricula.

The third theme was to revise our environmental programs and policies toward water, and to center them on the restoration of primeval populations of aquatic and marine animals as an important source of food, something that has been lost over the last three centuries. This new approach to the environment will have to be developed by private citizens grouped into watershed associations, operating free from industrial and commercial influences.

Underlying these three main themes is the need for our industrial society to learn about sustainable living styles from the indigenous societies of Native Americans that have successfully lived on the American continent for millennia. During the imperial expansion of European colonists into the Americas, these societies were largely destroyed. Fortunately, however, some indications of their ecological styles remain for us to study. We should enlist the help and wisdom of their remnant populations to guide us toward a sustainable future.

In the coming millennium, we must unite the original North and South American cultures, along with the cultures of the immigrant Americans from Europe, Africa, and Asia, to join together in a sustainable and just society of *new* Americans.

Changes

As a young engineer, I was taught to limit the scope of my analysis by isolating the object under study, and then drawing in only the forces acting on the isolated object. Also, I was taught to make simplifying assumptions in order to reduce the number of variables so that the equations could easily be solved. Many components of a problem had to be neglected in order to get a rapid solution.

But as an experienced planner and engineer, I now learn that those simplifications could be dangerous. It is not easy to change the approaches used since my youth. While the isolated objects and simplified assumptions may be adequate for a particle or for a beam, they are not adequate for bays and rivers, or for the real world.

We need to change the way we teach young engineers to build the *new* America.

American engineers and environmental planners have been designing and building dams, aqueducts, canals, and sewers for the last three centuries within a cultural context dominated by industrial corporations and a disregard for the sanctity of Creation. The primacy given by this industrial culture to manipulation of natural resources for profit and to obtaining quick returns on investments has led to short-sighted engineering designs that then produced unsustainable hydraulic systems. These faulty engineering systems resulted in severe toxic contamination of our environment, and accelerated the destruction of fish and mammal populations that were important elements of our food supply.

7.1 ECOLOGICAL SUCCESS OF TRADITIONAL AMERICAN SOCIETIES

Traditional societies, such as the original Americans who inhabited this continent before its colonization by Europeans, had a reverence for their environment that differed significantly from that of industrially and commercially dominated cultures. The communal, religious attitude toward nature of these aboriginal peoples allowed them to live in ecological harmony with nature in a sustainable pattern lasting about 10,000 years. Their sustainable lifestyles were developed since the last Ice Age, as the woodland ecology developed around them.

Another way of viewing the ecological success of the primeval Americans is that the groups which survived the 10,000 years were those that had sustainable attitudes toward nature, while those that abused their own environment eventually disappeared. One could think of this as competitive selection for sustainable cultures.

7.2 ECOLOGICAL FAILURE OF INDUSTRIAL SOCIETIES

In contrast with the primeval Americans, the industrial culture that dominated the continent at the end of the 20th Century nearly destroyed American ecology, eliminating large populations of fish and mammals. The forerunners of 20th-Century industrial and commercial corporations were the companies chartered by European kings to extract wealth from the Americas: the *Casa de Contratacion* designed to oversee the slave and silver trade by the Spaniards in South America, the piracy expeditions of Sir Francis Drake commissioned by Queen Elizabeth to steal the silver from the Spaniards under a British syndicate, and then the Hudson Bay Company created to harvest furs from North America. The Hudson Bay Company was the forerunner of the modern international corporation and is now the oldest existing company in the world. Its success gave rise to the Virginia Company, the Massachusetts Bay Company, and several others similarly formed for commercial profit. The primeval Americans could not protect themselves from these rapacious companies, and the continent quickly yielded to immigrant settlements and industry.

The primary ecologic damage inflicted by the chartered companies in the colonial era and their corporate descendants in the 19th Century was the over-harvesting of beaver and other small fur-bearing animals, the near extermination of the American buffalo, the organized slaughter of whales, and finally in the 20th Century, the elimination of pelagic fisheries off North America.

A slightly different group of corporations — the new manufacturing industries of the Industrial Revolution and their descendants — wreaked another kind of havoc on our environment: the discharge and discarding of all kinds of toxic and hazardous industrial wastes. Water pollution from the manufacture of paper, metal products, electrical products, pharmaceuticals, and textiles are a few of the most offensive industries.

7.3 CITIZEN PROTESTS

Starting around Boston, one of the first colonial settlements in America to severely contaminate its surroundings and deplete its primeval fisheries, citizen protests have periodically led to engineering and regulatory efforts at restoring the environment. However, these carefully engineered improvements were repeatedly overcome by population growth, expansion of industrial contamination, and rapacious commercial harvesting of fish and mammals. The fault was in the planning concepts embedded in the industrial cultures of these immigrant societies.

The most important of the periodic citizen outbursts were the Great Sanitary Awakening at the end of the 19th Century and the Green Movement at the end of the 20th Century. The Great Sanitary Awakening resulted from the development of the germ theory of disease. The Green Movement, which started on Earth Day 1970, resulted from gradual public appreciation of the concept of ecology, the interrelatedness of all Creation. The participation of private citizens in the restoration of our rivers and harbors may be the most important legacy of Earth Day.

7.4 DEATH OF THE GREEN MOVEMENT

The Green Movement was stopped in its tracks in the late 1990s because it was not politically prepared for the strong and well-financed reaction of large commercial and industrial corporations. These wealthy corporations bought the votes of the federal and state congresses in the U.S. and paid the legislators to reduce the budgets of the environmental agencies to a pittance. The corporations also lobbied heavily

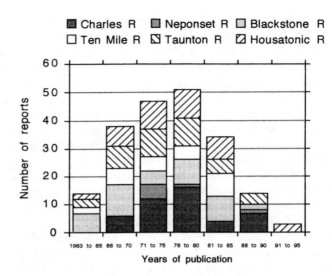

Figure 7.1   History of river surveys in Massachusetts. The activities of the river survey teams reached their highest point about 1980, declining rapidly thereafter as their budget and staff were slashed due to the conservative political movement to cut the environmental effort.

during the drafting of environmental regulations, which then made it almost impossible to force any of them to repair the ecological damage they had committed through the discharge and burying of toxic and hazardous wastes. By the end of the 20th Century, the Green Movement will no longer exist.

The impact of the movement to reduce government spending on the environment is clearly illustrated in a summary of the publications on water quality in six of the eastern river basins of Massachusetts, starting with the era just before Earth Day 1970 and continuing to the present (Figure 7.1). Before Earth Day, the Massachusetts Division of Water Pollution Control had a small team of scientists who conducted river surveys in the summer, producing about 12 reports in the 3 years from 1963 through 1965. As the Division was reorganized and strengthened, it gradually increased its staff of scientists and the monitoring of river and harbor conditions. By the last half of the 1970s, the Division was conducting and publishing over 50 river studies in 5 years, including river surveys, discharge surveys, basin planning documents, lake surveys, and coastal water studies.

Then the reaction set in, and large corporations called for reduced government spending. The budgets for environmental agencies were sharply slashed. Gradually, the field monitoring — the only true means of measuring environmental quality — had to be reduced as staff and budgets were cut. By the first half of the 1990s the number of river surveys was reduced to two in 5 years. From a staff of 40 scientists working on rivers in the late 1970s, the water quality group was reduced to three. Although the scientists of the Division compiled a valuable record on rivers, lakes, and harbors of Massachusetts during the Green Movement, the numbers clearly show that the Movement died in the 1990s.

7.5 CITIZEN WATERSHED ASSOCIATIONS

Although the Green Movement was overcome by large corporations, one legacy of the Green Movement may provide a key to eventually accomplishing its primary goal of permanently protecting our environment. This legacy is the formation of citizen-based watershed associations, aimed at the protection of entire river and estuary systems.

The beginning of the 21st Century would be an ideal time for these watershed associations to take away control of the rivers and oceans from the industrial interests that have predominated in America during the last three centuries. The value of riverside land and water rights has been so depressed by the last 300 years of industrial abuse that purchase of this land became fairly inexpensive at the end of the 20th Century. Thus, the watershed associations should position themselves to purchase riparian lands and then develop an economic base derived from sustainable use of these lands and water rights.

The associations should carefully organize their political base in order to develop as strong, self-financing organizations. Although the oldest and best-organized watershed associations may be in New England, some valuable concepts regarding self-sustainment can be learned from similar groups in the West. Revenue from increasingly productive fisheries, and from generation of hydroelectric power, should be used to sustain the associations.

Industries and corporations that had a major role in contamination of rivers and harbors, and in the destruction of aquatic and marine resources, might be willing to donate the riverside properties that they own, as a way of contributing to the restoration of these water bodies.

7.6 SUSTAINABLE PLANNING CONCEPTS

The nascent watershed associations, led by the most mature of the groups in New England, should adopt new and sustainable planning concepts for restoration of the rivers. Engineers and planners should use these concepts in the design of future dams, canals, aqueducts, and sewers. The concepts should be based on the reverential and sustainable attitudes of the traditional societies that endured for millennia before the industrial culture of the European immigrants overtook them.

Such traditional societies included the Algonkian culture of North America. Some of the important Algonkian cultural concepts were based on:

- A strongly democratic form of governing
- A reliance on consensus for decision-making
- Limited powers for head of state
- A profound understanding of plants and animals
- A reverence for all Creation
- The importance of traditional behavior based on solid principles that have a demonstrated survival value

7.7 INDUSTRIAL INTERFERENCE

The watershed associations must distance themselves from those industrial corporations that will try to use them to continue the exploitation of the river systems for corporate profit rather than for the general good. Perhaps the most delicate aspect of development of these new watershed associations will be for them to protect themselves from industrial interference. This interference may come directly as the industries try to purchase membership and influence, or indirectly through legislative restrictions. Unfortunately, many of the state and federal legislatures can be easily manipulated by large corporations.

7.8 NEW AMERICANS IN THE THIRD MILLENNIUM

The new American watershed associations should be gradually expanded to include coastal and ocean habitats. Under these associations, construction and rehabilitation of dams and sewers in the future should be based on new principles of sustainable design, solidly grounded in a historical perspective, rather than being based on the short-term economic analyses of the industrial past.

Sustainable design of dams, canals, and sewers must be based on principles, not on simplistic hydraulic concepts or narrowly conceived economics, or on thoughtless reactions to crises. A reverence for Creation or profound ecological awareness is the most important of these fundamental principles. The story of Noah's ark in the Hebrew scriptures reminds us that saving the human population is inextricably linked to preservation of all creatures.

> Then God said to Noah, "Go forth from the ark, you and your wife, and your sons and your sons' wives with you. Bring forth with you every living thing that is with you of all flesh — birds and animals and every creeping thing that creeps on the earth — that they may breed abundantly on the earth, and be fruitful and multiply upon the earth."

I think that those serious and open-minded planners and engineers who do not yet accept that our Universe is the handiwork of a Creator, would benefit from climbing a mountain at night, far from the smoke of the cities, and opening their eyes to look at the heavens. Or they might come closer to understanding by rising in the early morning to walk the forests and hear the birdsong, to watch the doe with her fawn along the wetlands, or to voyage to the sea and watch the whales jump for joy.

Unfortunately, many engineers have put their faith and future in supposedly rational but narrowly conceived planning procedures. These fads are enshrined in the engineering planning lexicon with such titles as: least-cost solutions, maximized benefit-cost ratios, maximum cost-effectiveness, linear programming, and near-optimal solutions.

The young engineers are usually taught to design dams based on these short-term economic analyses. Because of this seemingly rational but misdirected approach, their projects have depleted our riverine fisheries, caused erosion to destroy our estuaries, and depleted our coastlines of basic elements in the food chain for marine life. These designs could be considered rational only if reality were limited to short-term economic gain for a small group of investors. A much larger awareness of the beauty, fundamental importance to our life, and complexity of Creation is a basic necessity for any truly rational and sustainable design of water projects.

Sustainability is a trait of traditional societies, certainly, in part, because of its value in social evolution. Like species of plants or animals that exist today because their biological characteristics survived the hazards of the past better than those of other competing species, traditional societies survived because they developed behaviors and customs that protected the world around them better than did those of competing groups.

Those tribes that perfected weapons and developed strategies to kill all the deer in their forest in an orgy of hunting, eventually lost out to those tribes that saw the deer as fellow creatures. That is why the numerous and successful Plains tribes encountered by Wild Westerners in the Mississippi River Valley communed with the buffalo before hunting them, and hunted them carefully and reverently, only for necessities. These Plains tribes had persisted for 5000 to 10,000 years, developing with the Plains themselves. They were sustaining societies, surviving fierce and continual competition with their neighbors.

The traditions of these sustaining societies were not based on rational calculations of profit and costs, nor were they driven by hopes of short-term profits on ephemeral markets, but on reverence for Creation and its creatures. The apparent economic success of Wild Westerners in eliminating the buffalo and also of the Plains tribes should not deceive us. The dominance of the crusty New Englanders, Wild Westerners, and Spanish conquistadores over the rather passive populations they found in the Americas may also be a temporary illusion. The history of Wild Westerners is only about 200 years long; that of the New Englanders only about 400 years; and that of the conquistadores only about 500 years. Compare those short histories with the 10,000 years of evolution of the original American societies. The verdict on the immigrant societies should not be rendered for several thousand more years.

There have been some important successes in the Americas upon which we should build. The crusty New Englanders have learned to organize effective watershed associations, based on principles of the Iroquois and Algonkian peoples. The Wild Westerners have learned to "save the elk and deer" by protecting their habitats. The conquistadores have learned to save their rivers and sustain agriculture by sewage irrigation. Canadians have learned to design their hydroelectric dams with respect for the original Americans. Maybe all of us *new* Americans can combine these learning experiences.

The new watershed associations in parts of the Americas such as New England have been especially successful in those river basins where environmentalists and landowners have combined efforts with people dedicated to fishing, hunting, trapping, and genuine enjoyment of the outdoors. The importance of fishing and hunting groups in strong watershed associations is due to their understanding and appreciation of the important ecological role of fish and mammals, similar to the reverence for animals of the Algonkians.

It is very encouraging that the new American watershed associations have closely patterned themselves along Algonkian principles, based on admiration and love for our natural world as the supreme culmination of creation. Perhaps these groups are the forerunners of new Americans who can combine the best of the primeval traditions of the Algonkians, and the imported traditions of the Europeans, Africans, and Asians who have come to the continent more recently.

Epilogue

THE IMPORTANCE OF WATER

How essential are the oceans, rivers, mountains, and forests? For those who see this world as a miracle of a divine Creator, they are the basic media that give us life and in which we find our true humanity. Water represents the beginning of our life's journey. Water is at the core of life, especially in baptism for cleansing of sins and for renewal.

The Christian scriptures also remind us of the need for sustainable supplies of water. When a woman of Samaria questioned whether he was greater than the patriarch Jacob who gave them the well, Jesus reminded her that of the water from Jacob's well,

"Every one who drinks of this water will thirst again..."

LIST OF FIGURES

Cover: Bonneville Dam obstructs migrating salmon on the Columbia River in the Pacific Northwest.

LIST OF TABLES

SOURCES — MASSACHUSETTS LIBRARIES

Limited numbers of copies of those reports dealing with the waters of Massachusetts are available at no cost by written request to the Massachusetts Department of Environmental Protection at One Winter Street, Boston, MA 02108.

Furthermore, at the time of first printing, eight copies of each report were submitted to the State Library at the State House in Boston, which were subsequently distributed as follows:

- Two copies retained on the shelf at the State Library under the Golden Dome
- One microfilm copy retained at the State Library
- One copy to the Boston Public Library in Copley Square
- One copy to the Worcester Public Library
- One copy to the Springfield Public Library
- One copy to the University Library at the University of Massachusetts in Amherst
- One copy to the U.S. Library of Congress in Washington, D.C.

A complete list of reports published since 1963 is printed each July and also available at the Winter Street address above.

Index